The IMA Volumes in Mathematics and its Applications

Volume 146

Series Editors
Douglas N. Arnold   Arnd Scheel

Institute for Mathematics and its Applications (IMA)

The **Institute for Mathematics and its Applications** was established by a grant from the National Science Foundation to the University of Minnesota in 1982. The primary mission of the IMA is to foster research of a truly interdisciplinary nature, establishing links between mathematics of the highest caliber and important scientific and technological problems from other disciplines and industries. To this end, the IMA organizes a wide variety of programs, ranging from short intense workshops in areas of exceptional interest and opportunity to extensive thematic programs lasting a year. IMA Volumes are used to communicate results of these programs that we believe are of particular value to the broader scientific community.

The full list of IMA books can be found at the Web site of the Institute for Mathematics and its Applications:

http://www.ima.umn.edu/springer/volumes.html

Presentation materials from the IMA talks are available at

http://www.ima.umn.edu/talks/

Douglas N. Arnold, Director of the IMA

* * * * * * * * * *

IMA ANNUAL PROGRAMS

1982–1983	Statistical and Continuum Approaches to Phase Transition
1983–1984	Mathematical Models for the Economics of Decentralized Resource Allocation
1984–1985	Continuum Physics and Partial Differential Equations
1985–1986	Stochastic Differential Equations and Their Applications
1986–1987	Scientific Computation
1987–1988	Applied Combinatorics
1988–1989	Nonlinear Waves
1989–1990	Dynamical Systems and Their Applications
1990–1991	Phase Transitions and Free Boundaries
1991–1992	Applied Linear Algebra
1992–1993	Control Theory and its Applications
1993–1994	Emerging Applications of Probability
1994–1995	Waves and Scattering
1995–1996	Mathematical Methods in Material Science
1996–1997	Mathematics of High Performance Computing

Continued at the back

Alicia Dickenstein   Frank-Olaf Schreyer
Andrew J. Sommese
Editors

Algorithms in Algebraic Geometry

Alicia Dickenstein
Departamento de Matemática
Facultad de Ciencias Exactas y Naturales
Universidad de Buenos Aires
(1428) Buenos Aires
Argentina
http://mate.dm.uba.ar/~alidick/

Frank-Olaf Schreyer
Mathematik und Informatik
Universität des Saarlandes
Campus E2 4
D-66123 Saarbrücken
Germany
http://www.math.uni-sb.de/ag/schreyer/

Andrew J. Sommese
Department of Mathematics
University of Notre Dame
Notre Dame, IN 46556-4618
U.S.A.
http://www.nd.edu/~sommese/

Series Editors

ISBN 978-1-4419-2583-1
eISBN 978-0-387-75155-9

Mathematics Subject Classification (2000): Primary 11T71, 13P05, 14G05, 14H50, 14J70, 14M17, 14M99, 14P05, 14P99, 14Q05, 14Q10, 14Q99, 47B35, 52A20, 65H10, 65H20, 68Q25, 68W30, 90C22 Secondary 14M15

Camera-ready copy provided by the IMA.

9 8 7 6 5 4 3 2 1

springer.com

FOREWORD

This IMA Volume in Mathematics and its Applications

ALGORITHMS IN ALGEBRAIC GEOMETRY

contains papers presented at a highly successful one-week workshop on the same title. The event was an integral part of the 2006-2007 IMA Thematic Year on "Applications of Algebraic Geometry." We are grateful to all the participants for making this workshop a very productive and stimulating event. Special thanks to Alicia Dickenstein (Departamento de Matemática, University of Buenos Aires), Frank-Olaf Schreyer (Mathematik und Informatik, Universität des Saarlandes), and Andrew J. Sommese (Department of Mathematics, University of Notre Dame) for their superb role as workshop organizers and editors of these proceedings.

We take this opportunity to thank the National Science Foundation for its support of the IMA.

Series Editors

Douglas N. Arnold, Director of the IMA

Arnd Scheel, Deputy Director of the IMA

PREFACE

In the last decade, there has been a burgeoning of activity in the design and implementation of algorithms for algebraic geometric computation. Some of these algorithms were originally designed for abstract algebraic geometry, but now are of interest for use in applications and some of these algorithms were originally designed for applications, but now are of interest for use in abstract algebraic geometry.

The Workshop on *Algorithms in Algebraic Geometry* that was held in the framework of the IMA Annual Program Year in Applications of Algebraic Geometry by the Institute for Mathematics and Its Applications on September 18–22, 2006 at the University of Minnesota is one tangible indication of the interest. One hundred ten participants from eleven countries and twenty states came to listen to the many talks; discuss mathematics; and pursue collaborative work on the many faceted problems and the algorithms, both symbolic and numeric, that illuminate them.

This volume of articles captures some of the spirit of the IMA Workshop.

Daniel Bates, Chris Peterson, and Andrew Sommese show how the numerical algebraic geometry calculations originally aimed at applications may be used to quickly compute information about joins of varieties.

Frederic Bihan, J. Maurice Rojas, and Frank Sottile show the existence of fewnomial systems of polynomials whose number of positive real solutions equals a theoretical upper bound.

Sara Billey and Ravi Vakil, blend combinatorics and geometry, to give an algorithm for algebraically finding all flags in any zero-dimensional intersection of Schubert varieties with respect to any number of flags. This leads to a very easy method of checking that a structure constant for flag manifolds is zero.

Antonio Cafure, Guillermo Matera, and Ariel Waissbein study the problem of finding the inverse image of a point in the image of a rational map between vector spaces over finite fields. This problem is of great current interest in the coding community because it lies at the heart of a new approach to public key encryption in a world where it is ever more likely that quantum computers will allow quick factorization of integers and thereby dissolve the current encryption methods underlying secure transactions, e.g., by financial institutions.

Anton Leykin, Jan Verschelde, and Ailing Zhao study different approaches to the multiplicity structure of a singular isolated solution of a polynomial system, and, by so doing, present a "deflation method" based on higher order derivatives which effectively restores the quadratic convergence of Newton's method lost due to the singularity of the isolated solution.

Heidi Mork and Ragni Piene study polar and reciprocal varieties for possibly singular real algebraic curves. They show that in the case when only ordinary singularities are present on the curve, these associated varieties contain nonsingular points on all components of the original curve and can be used to investigate the components of the curve. They also give an example of a curve with nonordinary singularities for which this fails.

Jiawang Nie, Pablo Parrilo, and Bernd Sturmfels investigate the k-ellipse. This is the plane algebraic curve consisting of all points such that the sum of distances from k given points is a fixed number. They show how to write its defining polynomial equation as the determinant of a symmetric matrix of linear polynomials. Their representation extends to arbitrary dimensions, and it leads to new geometric applications of semidefinite programming.

Andrew J. Sommese, Jan Verschelde, and Charles Wampler present a numerical method (based on their diagonal homotopy algorithm for intersecting varieties) with the potential to allow solution of polynomial systems, which have relatively few solutions, but whose Bézout numbers are too large to allow solution by the usual homotopy continuation methods.

Other interesting subjects presented at the Workshop include the computation of the intersection and self-intersection loci of parameterized space algebraic surfaces and the study of projections methods for the topology of algebraic curves and surfaces, which are of interest in Computer Aided Geometric Design, algorithms for finding generators of rings of invariants of algebraic groups, the relation between the structure of Gröbner bases and the decomposition of polynomial systems and the complexity of Gröbner basis computations for regular and semi-regular systems, the factorization of sparse polynomials over number fields, counting rational points on varieties over finite fields, algorithms for mixed volume computation, the effectiveness of number theory in algebraic geometry, applications of monomial ideals and computational algebra in the reverse engineering of biological networks, the description of Newton polytopes of implicit equations using Tropical Geometry, the mathematical description of Maple's algebraic curves package and algorithms for finding all real solutions contained in a complex algebraic curve.

The interested reader can consult the talk materials and online videos of the lectures at
`http://www.ima.umn.edu/2006-2007/W9.18-22.06/abstracts.html`.

Alicia Dickenstein
Departamento de Matemática
Facultad de Ciencias Exactas y Naturales
Universidad de Buenos Aires
(1428) Buenos Aires
Argentina
http://mate.dm.uba.ar/~alidick/

Frank-Olaf Schreyer
Mathematik und Informatik
Universität des Saarlandes
Campus E2 4
D-66123 Saarbrücken
Germany
http://www.math.uni-sb.de/ag/schreyer/

Andrew J. Sommese
Department of Mathematics
University of Notre Dame
Notre Dame, IN 46556-4618
U.S.A.
http://www.nd.edu/~sommese/

CONTENTS

APPLICATION OF A NUMERICAL VERSION OF TERRACINI'S LEMMA FOR SECANTS AND JOINS

DANIEL J. BATES*, CHRIS PETERSON†, AND ANDREW J. SOMMESE‡

Abstract. This paper illustrates how methods such as homotopy continuation and monodromy, when combined with a numerical version of Terracini's lemma, can be used to produce a high probability algorithm for computing the dimensions of secant and join varieties. The use of numerical methods allows applications to problems that are difficult to handle by purely symbolic algorithms.

Key words. Generic point, witness point, homotopy continuation, irreducible components, numerical algebraic geometry, monodromy, polynomial system, secant, join.

AMS(MOS) subject classifications. Primary 65H10, 65H20, 68W30, 14Q99, 14M99.

1. Introduction. In this paper we study the dimension of secant and join varieties from a numerical point of view. In particular, we show how methods from numerical algebraic geometry can combine with a numerical version of Terracini's lemma to produce a high reliability algorithm for computing the dimension of such varieties. There are five primary situations where the utilization of numerical methods may be more advantageous than purely symbolic methods.

(1) The method can be applied to any subcollection of irreducible components of an algebraic set. In particular, it is not necessary to decompose the ideal defining the algebraic set before carrying out the computation.
(2) The ideals of the varieties involved in the computation can be generated by non-sparse polynomials with arbitrary (but bounded) coefficients. In other words, the coefficients can be any complex number; they are not restricted to be rational nor algebraic. Furthermore, it is not necessary to represent coefficients which are algebraic numbers as variables satisfying constraints. All that is needed is a good numerical approximation of the algebraic number.

*Institute for Mathematics and Its Applications (IMA), 400 Lind Hall, University of Minnesota, Minneapolis, MN 55455-0436; dabates@ima.umn.edu; www.ima.umn.edu/~dabates. The first author was supported by the Institute for Mathematics and Its Applications (IMA).

†Department of Mathematics, Colorado State University, Fort Collins, Colorado 80525; peterson@math.colostate.edu; www.math.colostate.edu/~peterson. The second author was supported by Colorado State University; NSF grant MSPA-MCS-0434351; and the Institute for Mathematics and Its Applications (IMA).

‡Department of Mathematics, University of Notre Dame, Notre Dame, IN 46556-4618; sommese@nd.edu; www.nd.edu/~sommese. The third author was supported by the Duncan Chair of the University of Notre Dame; the University of Notre Dame; NSF grant DMS-0410047; and the Institute for Mathematics and Its Applications (IMA).

This drastically reduces the computational complexity of algebraic systems with algebraic (or transcendental) coefficients.

(3) The varieties involved can have high codimension (provided the degrees of the generators of the ideal are not too large).

(4) The varieties are not required to have rational points.

(5) Information can be extracted even if the ideals utilized have generators with small inaccuracies in their coefficients.

Interest in secant and join varieties spans several fields including algebraic geometry, combinatorics, complexity theory, numerical analysis, statistics and physics [6, 8, 13, 23]. Much of this interest derives from the connection between secant varieties and problems involving tensor rank [5]. Rather than focus on the connection with tensor rank and optimizing in this direction, this paper will consider the general problem of computing the dimension of secant and join varieties. Examples are purposefully chosen to illustrate situations where a numeric-symbolic approach may be more natural than a symbolic approach alone. First we recall some basic definitions that will be used throughout the paper. If $Q_1, \dots, Q_p$ are points in $\mathbb{P}^m$ then we let $< Q_1, \dots, Q_p >$ denote their linear span. If $X_1, \dots, X_p$ are projective varieties in $\mathbb{P}^m$ then the *join* of the varieties, $J(X_1, \dots, X_p)$, is defined to be the Zariski closure of the union of the linear span of p-tuples of points $(Q_1, \dots, Q_p)$ where $Q_i \in X_i$. In other words

$$J(X_1, \dots, X_p) = \overline{\bigcup_{Q_1 \in X_1, \dots, Q_p \in X_p} < Q_1, \dots, Q_p >}.$$

If $X_1, \dots, X_p$ have dimensions $d_1, \dots, d_p$ then the expected dimension (and the maximum possible dimension) of $J(X_1, \dots, X_p)$ is $\min\{m, p-1+\sum d_i\}$. The *p-secant* variety of X is defined to be the join of p copies of X. We will denote this by $\sigma_p(X)$. For instance, $\sigma_2(X) = J(X, X)$ is the variety of secant lines to X. The expected dimension (and the maximum possible dimension) of $\sigma_p(X)$ is $\min\{m, pr + (p-1)\}$. X is said to have a *defective* p-secant variety if $\dim \sigma_p(X) < \min\{m, pr + (p-1)\}$ while X is called *defective* if there <u>exists</u> a p such that $\dim \sigma_p(X) < \min\{m, pr + (p-1)\}$.

Terracini's lemma is perhaps the most useful direct computational tool for computing the dimension of secant and join varieties. The lemma asserts that to compute the tangent space to a join variety, $J(X_1, \dots, X_s)$, at a generic point, P, it is enough to compute the span of the tangent spaces at generic points on each of the X_i's. Terracini's lemma was originally stated in the situation where $X_1 = \cdots = X_s$. This allowed one to compute the dimension of $\sigma_s(X_1)$ from s generic points on X_1. If, for points Q_i on X_i, the tangent spaces $T_{Q_1}, \dots, T_{Q_s}$ are independent then $J(X_1, \dots, X_s)$ has dimension $s - 1 + \sum_{i=1}^{s} \dim X_i$. This is precisely the dimension that $J(X_1, \dots, X_s)$ is expected to have in the situation where $m \geq s - 1 + \sum_{i=1}^{s} \dim X_i$. By upper semicontinuity, if there exist smooth points (not necessarily generic) such that the tangent spaces $T_{Q_1}, \dots, T_{Q_s}$

are independent (or else span the ambient space) then $J(X_1, \ldots, X_s)$ has the expected dimension. Thus, to show that X does not have a defective p-secant variety, it is enough to find p smooth points on X such that the tangent spaces at these points are either linearly independent or else span the ambient space. As a consequence, to check that X is not defective, it is enough to check that $\sigma_\alpha(X)$ and $\sigma_\beta(X)$ have the correct dimension when $\alpha = \max\{p \,|\, pr + p - 1 \leq m\}$ and $\beta = \min\{p \,|\, pr + p - 1 \geq m\}$.

Let $\mathbf{I}$ denote the homogeneous ideal of a variety V. One can apply Terracini's lemma to compute $\dim \sigma_p(V)$ by evaluating the Jacobian matrix of $\mathbf{I}$ at p general points on V and then utilizing each evaluated Jacobian matrix to construct a basis for the tangent spaces at these p general points. Since in practice one does not pick true general points, Terracini's lemma can be used to prove non-deficiency but it cannot be used to prove deficiency. Proving non-deficiency works extremely well for varieties where one can produce many exact points [2, 15]. However, the requirement of Terracini's lemma to understand the independence of tangent spaces at points on a variety poses, in general, an obstacle since *one cannot produce any exact points on a typical variety.* Since numerical techniques can be used to produce points *arbitrarily close* to a variety, one can hope to use a version of Terracini's lemma in the numerical setting.

The weakness of numerical methods, it is sometimes argued, is the loss of exactness. On the other hand, it is precisely this loss of exactness that allows numerical techniques to apply to problems that are unreachable by a purely symbolic algorithm. By combining numerical methods with ideas from symbolic computation, algebraic relationships can be made relatively stable under small perturbations and their presence can be detected. The main goal of this paper is to demonstrate a relatively elementary use of numerical-symbolic methods on a particular set of problems arising naturally in algebraic geometry. The method relies mainly on known theory, so the main contribution is the application of that theory rather than the theory itself.

2. Background.

2.1. Homotopy continuation. Given a polynomial system $F : \mathbb{C}^n \to \mathbb{C}^n$, one may compute accurate numerical approximations for all isolated solutions of F via homotopy continuation. This is a well-established numerical method in which F is cast as a member of a parameterized family of polynomial systems one of which, say G, has known isolated solutions. Given a homotopy such as $H = (1-t) \cdot F + \gamma \cdot t \cdot G$ where γ is a random complex number and G is a compatibly chosen polynomial system, there are paths leading from the isolated solutions of G at $t = 1$ to solutions of F at $t = 0$. These paths may be numerically tracked using standard predictor/corrector methods such as Euler's method and Newton's method. A collection of techniques, known as endgames, are utilized to reduce the computational cost of computing singular endpoints with a given reliability.

The set of path endpoints computed at $t = 0$ contains all isolated solutions of F as well as points lying on higher-dimensional irreducible components of the algebraic set defined by F.

Let $Dim(V)$ denote the dimension of the top dimensional component of the algebraic set, V, determined by F. The basic algorithms of numerical algebraic geometry produce discrete data called a *witness point set* [17, 20]. For each dimension d with $0 \leq d \leq Dim(V)$ this consists of a set of points W_d, called a *witness point set for dimension* d, and a generic codimension d linear space L_d with the basic property:

- Within a user-specified tolerance, the points of W_d are the intersections of L_d with the union of the d-dimensional components of V.

In the nonreduced case, there is some numerically useful extra information about the homotopy used to compute W_d. Since a general linear space meets each d-dimensional irreducible component W of V in exactly $Deg(W)$ points, each d-dimensional irreducible component W of V has at least one witness point in W_d. A *cascade algorithm* utilizing repeated applications of homotopy continuation to polynomial systems constructed from F yields the full witness set. Each of the polynomial systems constructed from F is obtained by adding extra linear equations (corresponding to slices by generic hyperplane sections).

Using techniques such as monodromy, it is possible to partition W_d into subsets, which are in one-to-one correspondence with the d-dimensional irreducible components of V. In particular, by tracking points in a witness set W_d around various loops in the parameter space, one can organize the points in W_d into sets such that all points of a set lie on the same irreducible component. Although a stopping criterion for monodromy by itself is unknown, linear traces [19] provide a means to certify, and sometimes even carry out, such partitions with certainty.

Thus, given an ideal $\mathbf{I}$, it is possible to produce by numerical methods a collection of subsets of points such that the subsets are in one to one correspondence with the irreducible components of the algebraic set determined by $\mathbf{I}$. Furthermore, the points within a given subset can be chosen to be within a pre-specified tolerance of the irreducible component represented by the subset. A classic introduction to general continuation methods can be found in [3]. For an overview of newer algorithms and techniques within this field, see [20, 21]. For details on the cascade algorithm, see [16, 20].

2.2. Singular value decomposition. Every $m \times n$ matrix M may be decomposed as $M = U\Sigma V^*$ where U and V are square, unitary matrices and Σ is an $m \times n$ diagonal matrix with real non-negative entries. This factorization is known as the singular value decomposition (SVD) of M. The diagonal entries of Σ are the singular values of M and the columns of U and V are the left and right singular vectors of M, respectively. A real number σ is a singular value if and only if there exist unit length vectors $\mathbf{u}, \mathbf{v}$ such that $M\mathbf{v} = \sigma\mathbf{u}$ and $M^*\mathbf{u} = \sigma\mathbf{v}$.

The singular value decomposition of a matrix is a key tool in determining both the numerical rank of a matrix and a basis for the nullspace of the matrix. In particular, for a matrix with exact entries, the number of nonzero singular values is exactly the rank of the matrix. In a floating point setting, one may compute the numerical rank of M by counting the number of singular values larger than some tolerance ϵ. Thus the numerical rank is a function of ϵ. This raises the natural question: *How do you choose ϵ?* A general precise answer to this question is both unclear and application-dependent. However, in the setting of numerical algebraic geometry it is often possible to increase certainty by increasing precision. Singular values which would be zero in an exact setting but which are non-zero due to imprecision and round off errors can be made to shrink towards zero by recomputing with increased precision. This will be illustrated in examples appearing later in the paper. Another fact that will be used is that the right singular vectors corresponding to the (numerically) zero singular values form a basis for the (numerical) nullspace of the matrix, as described in [25]. The SVD is also known for providing a method for producing the nearest rank k matrix to a given matrix (in the Frobenius norm), simply by setting to 0 the appropriate number of smallest entries of Σ and then re-multiplying.

The computation of the SVD of a matrix is more costly than the computation of the QR decomposition. However, the SVD yields critical information; may be stably computed; and is a good choice for ill-conditioned (i.e., nearly singular) matrices. The computation of the SVD relies on Householder reflectors or Givens rotations to reduce the matrix to bidiagonal form. From there, a number of methods related to the QR decomposition may be used to extract the full decomposition. A good general reference is provided by [26], while [22] gives a particularly direct description of the decomposition algorithm. It should be noted that many of the computations that are made in the following examples could also be computed using the rank revealing method of Li and Zeng [14]. However, for the purposes of this paper, the full singular value decomposition yielded more detailed information and could be efficiently computed. Currently, the SVD algorithm is the only rank revealing method implemented in Bertini.

2.3. Terracini's lemma. Terracini's lemma, as originally formulated, provides an effective method for computing the dimension of a secant variety [24]. The applicability of the lemma was later extended to join varieties and higher secant varieties [1]. The lemma asserts that the tangent space to a join variety, $J(X_1, \ldots, X_s)$, at a generic point, P, is equal to the span of the tangent spaces at related generic points on each of the X_i's. In particular, it states:

LEMMA 2.1 (**Terracini's lemma**). *Let $X_1, \ldots, X_s$ be irreducible varieties in $\mathbb{P}^n$ and let $Q_1, \ldots, Q_s$ be distinct generic points (with Q_i on X_i for each i). Let $T_{Q_i'}$ be the tangent space to the cone in $\mathbb{C}^{n+1}$ over X_i at a representative Q_i' of the point Q_i in $\mathbb{C}^{n+1} \setminus 0$; and let $\mathbf{P}(T_{Q_i})$ be the projectivize tangent space to X_i at Q_i in $\mathbb{P}^n$. Then* $\dim J(X_1, \ldots, X_s) = \dim < \mathbf{P}(T_{Q_1}), \ldots, \mathbf{P}(T_{Q_s}) > = \dim < T_{Q_1'}, \ldots, T_{Q_s'} > -1$.

By upper semicontinuity, it follows as an immediate corollary of Terracini's lemma that if there exist smooth points (not necessarily generic) such that the tangent spaces $T_{Q_1'}, \ldots, T_{Q_s'}$ are independent (or else span the ambient space) then $J(X_1, \ldots, X_s)$ has the expected dimension. Since one can never be sure of choosing true generic points, if the tangent spaces $T_{Q_1'}, \ldots, T_{Q_s'}$ are not independent, one cannot conclude that $J(X_1, \ldots, X_s)$ does not have the expected dimension. Thus, while Terracini's lemma can be used to furnish proofs in the nondefective case, it can only be used as a *guide* in the defective case.

2.4. Bertini software package. All examples of this paper were run using Bertini and Maple™. Bertini [4] is a software package under continuing development for computation in numerical algebraic geometry. Bertini is written in the C programming language and makes use of the GMP-based library MPFR for multiple precision floating point numbers. It also makes use of adaptive multiprecision; automatic differentiation; and straight line programs for efficient evaluation. Bertini will accept as input a set of homogeneous polynomials over any product of projective spaces or a set of inhomogeneous polynomials over any product of complex spaces.

Bertini is capable of finding all complex solutions of a polynomial system with complex coefficients via methods rooted in homotopy continuation. By using automatic adaptive precision techniques and special methods known as endgames, Bertini is capable of providing these solutions to several thousand digits.

3. Five illustrative examples. In the first subsection below, we illustrate an exact method and two approximate methods in conjunction with Terracini's lemma applied to the Veronese surface in $\mathbb{P}^5$. The second subsection contains two examples chosen to illustrate situations well-suited to a numeric approach but that may be difficult via Gröbner basis techniques. There are certainly situations where a symbolic approach is preferable to a numeric approach. For instance, it is easy to produce exact generic points on a parameterized variety and to determine the tangent spaces at these points. Furthermore, this can typically be done in such a manner that modular techniques can be utilized. However, for a typical non-parameterized variety, exact generic points can't be produced. As a consequence, the methods illustrated in this paper should be seen as complementary to the exact methods that one can employ using programs such as [7, 11, 12]. Perhaps in the future, symbolic based programs will be able to exchange information in a meaningful way with numeric based programs.

3.1. Secant variety of the Veronese surface in $\mathbb{P}^5$. In this section we show, in some detail, three methods utilizing Terracini's Lemma to compute the dimension of the secant variety to a variety. For simplicity we use the well-studied Veronese surface in $\mathbb{P}^5$ to illustrate each approach. In each of the three methods, we compute an (approximate) basis for the tangent space to the affine cone of the variety at two distinct (approximate) points P_1, P_2. We next look at the dimension of the span of the two tangent spaces and then apply Terracini's Lemma to interpret the results. In Example 1, we review the standard method to apply Terracini's Lemma. In Example 2, we modify the standard approach to fit a numerical setting. Finally, in Example 3 we use a numerical version of Terracini's lemma in the setting of a highly sampled variety.

EXAMPLE 1 (*Secant variety by the exact Jacobian method*). Let $\mathbf{I} = (F_1, \ldots, F_t) \subseteq \mathbb{C}[x_1 \ldots, x_n]$. The associated Jacobian matrix of $\mathbf{I}$ is the $t \times n$ matrix whose $(i,j)^{th}$ entry is $\frac{\partial F_i}{\partial x_j}$. The Veronese surface $V \subseteq \mathbb{P}^5$ is defined as the image of the map $[x : y : z] \to [x^2 : xy : xz : y^2 : yz : z^2]$. The homogeneous ideal of V is $\mathbf{I}_V = (e^2 - df, ce - bf, cd - be, c^2 - af, bc - ae, b^2 - ad)$. The associated Jacobian matrix of $\mathbf{I}_V$ is therefore

$$Jac(a,b,c,d,e,f) = \begin{pmatrix} 0 & 0 & 0 & -f & 2e & -d \\ 0 & -f & e & 0 & c & -b \\ 0 & -e & d & c & -b & 0 \\ -f & 0 & 2c & 0 & 0 & -a \\ -e & c & b & 0 & -a & 0 \\ -d & 2b & 0 & -a & 0 & 0 \end{pmatrix}.$$

Consider the points $[1:3:7]$ and $[1:5:11]$ in $\mathbb{P}^2$. They map to the points $P = [1:3:7:9:21:49]$ and $Q = [1:5:11:25:55:121]$ in $\mathbb{P}^5$. Evaluating the Jacobian matrix at P yields the matrix

$$\begin{pmatrix} 0 & 0 & 0 & -49 & 42 & -9 \\ 0 & -49 & 21 & 0 & 7 & -3 \\ 0 & -21 & 9 & 7 & -3 & 0 \\ -49 & 0 & 14 & 0 & 0 & -1 \\ -21 & 7 & 3 & 0 & -1 & 0 \\ -9 & 6 & 0 & -1 & 0 & 0 \end{pmatrix}.$$

This matrix has rank 3 and spans the normal space to V at P. The tangent space T_P to V at P (as an affine variety) is spanned by the right null vectors to this matrix, thus T_P is the row space of the matrix

$$A = \begin{pmatrix} 2 & 3 & 7 & 0 & 0 & 0 \\ 0 & 1 & 0 & 6 & 7 & 0 \\ -1 & -3 & 0 & -9 & 0 & 49 \end{pmatrix}.$$

Similarly, we determine that T_Q is the row space of the matrix

$$B = \begin{pmatrix} 2 & 5 & 11 & 0 & 0 & 0 \\ 0 & 1 & 0 & 10 & 11 & 0 \\ -1 & -5 & 0 & -25 & 0 & 121 \end{pmatrix}.$$

The dimension of the space spanned by the row spaces of A and B is, of course, the rank of the matrix

$$C = \begin{pmatrix} 2 & 3 & 7 & 0 & 0 & 0 \\ 0 & 1 & 0 & 6 & 7 & 0 \\ -1 & -3 & 0 & -9 & 0 & 49 \\ 2 & 5 & 11 & 0 & 0 & 0 \\ 0 & 1 & 0 & 10 & 11 & 0 \\ -1 & -5 & 0 & -25 & 0 & 121 \end{pmatrix}.$$

A quick computation (in exact arithmetic) shows the rank of C to be 5. As a consequence, the secant variety to the Veronese surface is (most likely) four dimensional as a projective variety.

It is important to note that the example above does not give a proof that the dimension of the secant variety is four since it is possible (though unlikely) that the points we chose were special points for this particular example. Terracini's Lemma requires the points to be generic. If the rank of the matrix had been 6 then we would have a proof that the secant variety is five dimensional since the points were smooth points on the surface, the dimension achieved the maximum possible and we could apply upper semi-continuity. In the next two examples, we illustrate two numerical methods for "determining" the dimension of the secant variety to the Veronese.

EXAMPLE 2 (*Secant variety by the approximate Jacobian method*). Consider the points $\left[\sqrt{2} : \frac{22}{7} : 3^{(1/7)}\right]$ and $\left[7^{(1/5)} : \frac{15}{2} : \sqrt{5}\right]$. As in the example above, let P and Q be their images in $\mathbb{P}^5$. Let P' and Q' denote the points obtained from P, Q by rounding their coordinates to 40 digits of accuracy. The Jacobian matrix evaluated at P' has 3 singular values between 12 and 15 and 3 singular values between 10^{-42} and 10^{-38}. The Jacobian matrix evaluated at Q' has 3 singular values between 61 and 68 and 3 between 10^{-42} and 10^{-38}. The numerical rank of each matrix is 3 provided we round "appropriately small" singular values to zero. We use the singular value decomposition to find the closest rank 3 matrices to the Jacobians of P' and Q' and to find 3 linearly independent vectors in the right null space of each of these rank 3 matrices. We use these 2 sets of three vectors to build a 6×6 matrix, C. Finally, we apply the singular value decomposition to C and find five singular values between 0.1 and 1.5 and one small singular value that is approximately 2.8×10^{-40}. Thus we have obtained a matrix that, in an appropriate sense, has a numerical rank equal to 5. Again this suggests that the secant variety to the Veronese surface is (most likely) four dimensional as a projective variety.

EXAMPLE 3 (*Secant variety by a sampling method*). In this example we start by constructing two small clusters of 5 points each on the Veronese surface. We use these clusters to approximate the tangent space to the cone of the surface at the points corresponding to the centers of the clusters. In particular we chose the points $[0.2 : 0.3 : 0.4]$ and $[0.8 : 0.6 : 0.4]$ and then varied each of them by 0.0001 or 0.0002 in 5 (pseudo)random directions. By looking at their images in $\mathbb{C}^6$ we obtained two different 5×6 matrices, A and B. The 5 singular values of each of these matrices were on the order of $1, 10^{-4}, 10^{-4}, 10^{-8}, 10^{-8}$. Knowing ahead of time that we were trying to approximate the tangent space to the cone of a surface, we took the best rank 3 matrix approximations of A and B (still 5×6 matrices) and interpreted the row spaces of these matrices as representing the tangent spaces. Stacking these rank 3 matrices yielded a 10×6 matrix which had 5 singular values between 0.3 and 1.5 and one singular value in the range of 4.9×10^{-6}. Thus we obtained a matrix that, within an appropriate tolerance, has a numerical rank equal to 5. This suggests, once again, that the secant variety to the Veronese surface is four dimensional as a projective variety. Repeating the experiment with perturbations on the order of 0.0000015 yielded a matrix whose smallest singular value was roughly 5×10^{-9}.

Each of the previous three examples gives consistent and strong evidence for the dimension of a certain secant variety. None of the computations *prove* that this dimension is correct. If a variety is nondefective then exact arithmetic can typically be used to prove this fact *provided sufficiently general points can be found which lie on the variety.* Unfortunately, it is usually impossible to find any points which lie on a given variety. This necessitates either a different symbolic approach or else a non-symbolic approach. The numeric approach presented in this paper offers one possibility.

3.2. Two more examples. The following two examples present situations that may well frustrate a symbolic algebra system due to computational complexity. It is meant to illustrate a situation well suited to a numerical approach but not well suited to a symbolic approach.

EXAMPLE 4 (*Three secant variety of a curve in* $\mathbb{P}^8$). In this example, we let C be a curve in $\mathbb{P}^8$ defined as the complete intersection of 6 quadrics and a cubic. The coefficients of the defining equations were chosen randomly in the interval $[-3, 3]$ and included 16 digits past the decimal point. We used the homotopy continuation implementation in *Bertini* to produce three points whose coordinates are within 10^{-8} of an exact point on C. Performing computations with 16 digits of accuracy, we used the SVD to compute approximate bases for the tangent spaces to C (as an affine cone) for each of the three points. We combined the approximated basis vectors into a single 6×9 matrix. The singular value decomposition determined that the matrix had 6 positive singular values ranging from 0.35 to 1.37.

The computation was repeated at 50 digits of accuracy and the smallest singular value was again roughly 0.35. The higher accuracy computation caused the smallest singular value to move by less than 10^{-11}. As a result, we confirm the fact (known by other methods) that $\sigma_3(C)$ is 5 dimensional as a projective variety.

EXAMPLE 5 (*Jumbled Segre Variety*). In this example, we took the Segre embedding of $\mathbb{P}^2 \times \mathbb{P}^2$ into $\mathbb{P}^8$. Next a random linear change of variables was performed in order to produce a Segre variety, S, whose homogeneous ideal had a scrambled set of generators. The linear change of variables that we used involved various complicated algebraic and transcendental numbers. We used homotopy continuation in Bertini to produce two random points with coordinates within 10^{-8} of the coordinates of an exact point on S as an affine variety. The singular value decomposition was applied to find bases for the tangent space to S at the two points. The basis vectors for the tangent spaces were combined to form a 10×9 matrix. The SVD was applied to determine that the matrix had 8 singular values between 0.36 and 1.42 and 1 singular value that was approximately 0.91×10^{-11}. This result suggests that the dimension of the secant variety to the Segre variety is 7. In other words, that the Segre variety is 2-deficient. This deficiency is well known by other methods.

4. Further computational experiments.

4.1. Computations on systems which are close together.

Many symbolic algorithms require a computation of a Gröbner basis. If the defining equations of an ideal involve many different prime numbers, then the rational operations involved in the computation of the Gröbner basis can lead to fractions whose numerator and denominator are extremely large. In other words, symbolic techniques over $\mathbb{Q}$ often lead to *coefficient blowup.* In numerical methods, coefficient blowup of this type is avoided by employing finite precision arithmetic at various parts of a calculation. However, the small inaccuracies that occur when using finite precision can accumulate. As a consequence, it is crucial that the nature of these errors is understood in order to ensure that meaning can be attached to the output of the computation. A symbolic version of roundoff that is frequently utilized, to prevent coefficient blowup, is to carry out computations with exact arithmetic but over a field with finite characteristic. This can lead to problems. There are standard methods employed to increase the chance of a proper interpretation of the output. For instance, the computation can be made over a field with a sufficiently large characteristic or the computation can be repeated over fields with differing characteristic, etc. In the computer package Bertini, adaptive multiprecision is employed in a similar manner to help avoid potential pitfalls of low accuracy computations. The following example describes how computations will vary on two slightly different systems of equations. The first part of the example is carried out using an exact system of equations. The second part of the example is carried out after the system has been slightly perturbed.

EXAMPLE 6 (*The twisted cubic and the perturbed twisted cubic*). In this example, we start with the twisted cubic defined by the ideal $\mathbf{I} = (x^2 - wy, xy - wz, y^2 - xz)$. We can compute basis vectors for the tangent space to the twisted cubic at two exact points. Upon combining these vectors into a single 4×4 matrix we find that the matrix has rank 4 using exact arithmetic. This leads us to conclude that the secant variety of the twisted cubic fills the ambient space. If we repeat the computation with inexact points that lie very close to the twisted cubic (as an affine variety) then we obtain a 4×4 matrix which has no minuscule singular values. This leads us to suspect that the secant variety fills the ambient space. Now suppose we perturb each of the 3 quadric generators of $\mathbf{I}$ via a random homogeneous quadric with very small coefficients. The perturbed ideal defines 8 distinct points. Assuming the perturbation is very small, the homotopy continuation algorithm and the cascade algorithm will conclude that the perturbed ideal defines a curve if one chooses a certain range of tolerances but will conclude that the ideal defines 8 points if the tolerance is chosen sufficiently small.

4.2. Determining equations from a generic point. Let $\mathbf{I} \subseteq \mathbb{C}[x_0, \ldots, x_n]$ be the homogenous ideal of a projective variety V. Let $\mathbf{J}$ be a homogeneous ideal whose saturation with respect to the maximal ideal is $\mathbf{I}$. Suppose there exists a set of generators for $\mathbf{J}$ lying in $\mathbb{Q}[x_0, \ldots, x_n]$. Let $Sec(\mathbf{I})$ denote the homogenous ideal of the secant variety of V. Since Gröbner basis algorithms can be used to determine $\mathbf{I}$ from $\mathbf{J}$ and $Sec(\mathbf{I})$ from $\mathbf{I}$, there must exist a set of generators for $Sec(\mathbf{I})$ which lie in $\mathbb{Q}[x_0, \ldots, x_n]$.

Using the methods described in the previous sections, one can produce points arbitrarily close to generic points on V. By taking a general point on the secant line through two such points, one can produce points arbitrarily close to generic points on $Sec(V)$. The next example illustrates that from a numerical approximation to a single generic point on a variety, one *may* be able to obtain exact equations in the homogeneous ideal of the variety.

EXAMPLE 7. In this example we consider the Veronese surface $S \subset \mathbb{P}^5$ in its standard embedding. The homogeneous ideal of S has a generating set lying in $\mathbb{Q}[a, \ldots, f]$. Thus the homogeneous ideal of the secant variety to the Veronese surface must lie in $\mathbb{Q}[a, \ldots, f]$. We took two points on the Veronese surface involving a combination of algebraic and transcendental numbers. We then took a weighted sum of the points to obtain a point, P, on the secant variety of the Veronese surface. We next took a decimal approximation of the coordinates accurate to 75 digits to obtain a point Q which lies very close to the secant variety. The LLL and PSLQ algorithms provide a means for finding vectors, with relatively small integral entries, whose dot product with a given vector is very close to zero. We decided to use the PSLQ algorithm (as implemented in Maple™) due to its claimed increased numerical stability [9, 10]. We applied the algorithm to the 3-uple embedding of Q (a vector in $\mathbb{R}^{56}$). The outcome was a vector (almost) in the null space of the 3-uple embedding which corresponded to the equation $adf - ae^2 + 2bce - b^2f - c^2d \in \mathbb{Q}[a, \dots, f]$. The vanishing locus of this equation is well known as the hypersurface corresponding to the secant variety of S.

5. Conclusions. An application of a numeric-symbolic algorithm has been presented which allows estimates of the dimension of join and secant varieties to be made *with high reliability*. The examples were run in the Bertini software package [4] with some additional computations run in Maple™. Evidence of the defectivity of the Veronese surface in $\mathbb{P}^5$ was provided via three different methods (all using Terracini's lemma). Other examples were presented which demonstrate situations well suited to numerical methods but not well suited to exact methods. An example of a perturbed ideal was included to illustrate that care must be used when applying numerical techniques and that increased precision can be used to increase reliability of output. The example of the perturbed twisted cubic simultaneously demonstrates a strength and a limitation. The perturbed system is, in a numerical sense, very close to the original system. However, in an exact sense, it describes a variety with completely different characteristics. A symbolic algorithm would certainly distinguish between these two examples while a numerical-symbolic system only differentiates between the two varieties when very high precision is utilized. With a numerical approach, the closeness of the perturbed ideal to an ideal of a curve can be detected. Thus the loss of exactness, that comes with a numerical approach, is accompanied by a new collection of questions that can be answered and new regions of problems from which meaningful information may be extracted.

One could improve the approach of this paper by developing an array of criteria for certifying that a small singular value would indeed converge to zero if infinite precision could be utilized. The ultimate cost of such a certification would likely be very high. When determining which singular values of a matrix are zero via numerical methods, we were forced to choose a threshold below which a singular value is deemed to be zero. If one is dealing with inexact equations then there is a limit on the accuracy of the information obtained. If one can quantify the inaccuracy then reasonable thresholds can be determined. However, if one is dealing with exact equations, increased confidence in the result of a computation follows from performing the computation at a higher level of precision. This approach was used in Example 5 and is satisfactory in many, perhaps even most, situations. However, there will always be problems that can be constructed which can fool a numerically based system. By starting with exact equations and then using numerical methods one is necessarily losing information. It is an important research problem to quantify this in such a way as to yield improved confidence levels.

The authors would like to thank the referees of this paper for their comments and suggestions. Most of their comments and suggestions were incorporated and helped to improve the paper.

REFERENCES

[1] B. Ådlandsvik. *Joins and higher secant varieties.* Math. Scand. **61** (1987), no. 2, 213–222.

[2] H. Abo, G. Ottaviani, and C. Peterson. Induction for secant varieties of Segre varieties. Preprint: math.AG/0607191.

[3] E. Allgower and K. Georg. *Introduction to numerical continuation methods.* Classics in Applied Mathematics 45, SIAM Press, Philadelphia, 2003.

[4] D.J. Bates, J.D. Hauenstein, A.J Sommese, and C.W. Wampler. Bertini: Software for Numerical Algebraic Geometry. Available at www.nd.edu/~sommese/bertini.

[5] M.V. Catalisano, A.V. Geramita, and A. Gimigliano. *Ranks of tensors, secant varieties of Segre varieties and fat points.* Linear Algebra Appl. **355** (2002), 263–285.

[6] C. Ciliberto. *Geometric aspects of polynomial interpolation in more variables and of Waring's problem.* European Congress of Mathematics, Vol. I (Barcelona, 2000), 289–316, Progr. Math., **201**, Birkhäuser, Basel, 2001.

[7] CoCoATeam, CoCoA: *a system for doing Computations in Commutative Algebra.* Available at www.cocoa.dima.unige.it.

[8] J. Draisma. *A tropical approach to secant dimensions,* math.AG/0605345.

[9] H.R.P. Ferguson and D.H. Bailey. *A Polynomial Time, Numerically Stable Integer Relation Algorithm.* RNR Techn. Rept. RNR-91-032, Jul. 14, 1992.

[10] H.R.P. Ferguson, D.H. Bailey, and S. Arno. *Analysis of PSLQ, An Integer Relation Finding Algorithm.* Math. Comput. **68**, 351–369, 1999.

[11] D. Grayson and M. Stillman. Macaulay 2: *a software system for research in algebraic geometry.* Available at www.math.uiuc.edu/Macaulay2.

[12] G.M. Greuel, G. Pfister, and H. Schönemann. Singular 3.0: *A Computer Algebra System for Polynomial Computations.* Available at www.singular.uni-kl.de.

[13] J.M. Landsberg. *The border rank of the multiplication of two by two matrices is seven.*, J. Amer. Math. Soc. **19** (2006), 447–459.

[14] T.Y. Li and Z. Zeng. *A rank-revealing method with updating, downdating and applications.* SIAM J. Matrix Anal. Appl. **26** (2005), 918–946.

[15] B. McGillivray. *A probabilistic algorithm for the secant defect of Grassmann varieties.* Linear Algebra and its Applications **418** (2006), 708–718.

[16] A.J. Sommese and J. Verschelde. *Numerical Homotopies to compute generic points on positive dimensional Algebraic Sets.* Journal of Complexity **16**(3): 572–602 (2000),

[17] A.J. Sommese, J. Verschelde, and C.W. Wampler. *Numerical decomposition of the solution sets of polynomials into irreducible components.* SIAM J. Numer. Anal. **38** (2001), 2022–2046.

[18] A.J. Sommese, J. Verschelde, and C.W. Wampler. *A method for tracking singular paths with application to the numerical irreducible decomposition. Algebraic Geometry, a Volume in Memory of Paolo Francia.* Ed. by M.C. Beltrametti, F. Catanese, C. Ciliberto, A. Lanteri, and C. Pedrini. De Gruyter, Berlin, 2002, 329–345.

[19] A.J. SOMMESE, J. VERSCHELDE, AND C.W. WAMPLER. *Symmetric functions applied to decomposing solution sets of polynomial systems.* SIAM Journal on Numerical Analysis **40** (2002), 2026–2046.

[20] A.J. SOMMESE AND C.W. WAMPLER. *The Numerical Solution to Systems of Polynomials Arising in Engineering and Science.* World Scientific, Singapore, 2005.

[21] H. STETTER. *Numerical Polynomial Algebra.* SIAM Press, Philadelphia, 2004.

[22] G.W. STEWART. *Matrix Algorithms 1: Basic Decompositions.* SIAM Press, Philadelphia, 1998.

[23] B. STURMFELS AND S. SULLIVANT. *Combinatorial secant varieties.* Pure Appl. Math. Q. **2** (2006), 867–891.

[24] A. TERRACINI. *Sulla rappresentazione delle forme quaternarie mediante somme di potenze di forme lineari.* Atti R. Accad. delle Scienze di Torino, Vol. **51**, 1915–16.

[25] L. TREFETHEN AND D. BAU. *Numerical Linear Algebra.* SIAM Press, Philadelphia, 1997.

[26] G.H. GOLUB AND C.F. VAN LOAN. *Matrix computations*, 3rd edition. Johns Hopkins University Press, Baltimore, MD, 1996.

ON THE SHARPNESS OF FEWNOMIAL BOUNDS AND THE NUMBER OF COMPONENTS OF FEWNOMIAL HYPERSURFACES

FRÉDÉRIC BIHAN*, J. MAURICE ROJAS†, AND FRANK SOTTILE‡

Abstract. We prove the existence of systems of n polynomial equations in n variables with a total of $n+k+1$ distinct monomial terms possessing $\left\lfloor\frac{n+k}{k}\right\rfloor^k$ nondegenerate positive solutions. This shows that the recent upper bound of $\frac{e^2+3}{4}2^{\binom{k}{2}}n^k$ for the number of nondegenerate positive solutions has the correct order for fixed k and large n. We also adapt a method of Perrucci to show that there are fewer than $\frac{e^2+3}{4}2^{\binom{k+1}{2}}2^n n^{k+1}$ connected components in a smooth hypersurface in the positive orthant of $\mathbb{R}^N$ defined by a polynomial with $n+k+1$ monomials, where n is the dimension of the affine span of the exponent vectors. Our results hold for polynomials with real exponents.

Key words. Fewnomials, connected component.

AMS(MOS) subject classifications. 14P99.

1. Introduction. Khovanskii's Theorem on Real Fewnomials [5] implies that there are at most $2^{\binom{n+k}{2}}(n+1)^{n+k}$ nondegenerate positive solutions to a system of n polynomial equations in n variables which are linear combinations of (the same) $n+k+1$ monomials. This fewnomial bound is also valid in the more general setting of linear combinations of monomials with real-number exponents. The underlying bounds are identical whether one uses integral or real exponents [8], and the arguments of [3] require that we allow real exponents.

While Khovanskii's fewnomial bound was not believed to be sharp, only recently have smaller bounds been found. The first breakthrough was due to Li, Rojas, and Wang [7] who showed that a system of two trinomials in two variables has at most 5 positive solutions — which is smaller than Khovanskii's bound of 5184. Bihan [2] showed that a system of n polynomials in n variables with $n+2$ monomials has at most $n+1$ nondegenerate positive solutions and proved the existence of such a system with $n+1$ positive solutions. Bihan and Sottile [3] generalized this to all k, giving the upper bound of $\frac{e^2+3}{4}2^{\binom{k}{2}}n^k$ for the number of nondegenerate positive solutions, which is significantly smaller than Khovanskii's bound.

*Laboratoire de Mathématiques, Université de Savoie, 73376 Le Bourget-du-Lac Cedex, France (Frederic.Bihan@univ-savoie.fr, www.lama.univ-savoie.fr/~bihan).

†Department of Mathematics, Texas A&M University, College Station, Texas 77843-3368, USA (rojas@math.tamu.edu, www.math.tamu.edu/~rojas). Rojas supported by NSF CAREER grant DMS-0349309.

‡Department of Mathematics, Texas A&M University, College Station, Texas 77843-3368, USA (sottile@math.tamu.edu, www.math.tamu.edu/~sottile). Sottile supported by NSF CAREER grant DMS-0538734.

1.1. A lower bound for fewnomial systems. We show that the Bihan-Sottile upper bound [3] is near-optimal for fixed k and large n.

THEOREM 1. *For any positive integers n, k with $n > k$, there exists a system of n polynomials in n variables involving $n+k+1$ distinct monomials and having $\lfloor \frac{n+k}{k} \rfloor^k$ nondegenerate positive solutions.*
We believe that there is room for improvement in the dependence on k, both in the upper bound of [3] and in the lower bound of Theorem 1.

Proof. We will construct such a system when $n = km$, a multiple of k, from which we may deduce the general case as follows. Suppose that $n = km + j$ with $1 \leq j < k$ and we have a system of mk equations in mk variables involving $mk+k+1$ monomials with $(m+1)^k$ nondegenerate positive solutions. We add j new variables $x_1, \ldots, x_j$ and j new equations $x_1 = 1, \ldots, x_j = 1$. Since the polynomials in the original system may be assumed to have constant terms, this gives a system with n polynomials in n variables having $n+k+1$ monomials and $(m+1)^k = \lfloor \frac{n+k}{k} \rfloor^k$ nondegenerate positive solutions. So let us fix positive integers k, m and set $n = km$.

Bihan [2] showed there exists a system of m polynomials in m variables

$$f_1(y_1, \ldots, y_m) = \cdots = f_m(y_1, \ldots, y_m) = 0$$

having $m+1$ solutions, and where each polynomial has the same $m+2$ monomials, one of which we may take to be a constant.

For each $j = 1, \ldots, k$, let $y_{j,1}, \ldots, y_{j,m}$ be m new variables and consider the system

$$f_1(y_{j,1}, \ldots, y_{j,m}) = \cdots = f_m(y_{j,1}, \ldots, y_{j,m}) = 0,$$

which has $m+1$ positive solutions in $(y_{j,1}, \ldots, y_{j,m})$. As the sets of variables are disjoint, the combined system consisting of all km polynomials in all km variables has $(m+1)^k$ positive solutions. Each subsystem has $m+2$ monomials, one of which is a constant. Thus the combined system has $1 + k(m+1) = km+k+1 = n+k+1$ monomials. □

REMARK 1.1. Our proof of Theorem 1 is based on Bihan's nonconstructive proof of the existence of a system of n polynomials in n variables having $n+2$ monomials and $n+1$ nondegenerate positive solutions. While finding such systems explicitly is challenging in general, let us do so for $n \in \{2, 3\}$.

The system of $n = 2$ equations with 2 variables

$$x^2y - (1 + 4x^2) = xy - (4 + x^2) = 0,$$

has $4 = 2+1+1$ monomials and exactly 3 complex solutions, each of which is nondegenerate and lies in the positive quadrant. We give numerical approximations, computed with the computer algebra system SINGULAR [4, 10],

$$(2.618034, 4.1459), \quad (1, 5), \quad (0.381966, 10.854102).$$

The system of $n = 3$ equations with 3 variables

$$yz - (1/8 + 2x^2) = xy - (1/220 + x^2) = z - (1 + x^2) = 0,$$

has $5 = 3+1+1$ monomials and exactly 4 complex solutions, each of which is nondegenerate and lies in the positive octant,

$$(0.076645, 0.1359, 1.00587), \quad (0.084513, 0.13829, 1.00714),$$
$$(0.54046, 0.54887, 1.2921), \quad (1.29838, 1.30188, 2.6858).$$

1.2. An upper bound for fewnomial hypersurfaces. Khovanskii also considered smooth hypersurfaces in the positive orthant $\mathbb{R}^n_>$ defined by polynomials with $n+k+1$ monomials. He showed [6, Sec. 3.14, Cor. 4] that the total Betti number of such a fewnomial hypersurface is at most

$$(2n^2 - n + 1)^{n+k}(2n)^{n-1}2^{\binom{n+k}{2}}.$$

Li, Rojas, and Wang [7] bounded the number of connected components of such a hypersurface by $n(n+1)^{n+k+1}2^{n-1}2^{\binom{n+k+1}{2}}$. Perrucci [9] lowered this bound to $(n+1)^{n+k}2^{1+\binom{n+k}{2}}$. His method was to bound the number of compact components and then use an argument based on the faces of the n-dimensional cube to bound the number of all components. We improve Perrucci's method, using the n-simplex and the bounds of Bihan and Sottile [3] to obtain a new, lower bound.

THEOREM 2. *A smooth hypersurface in $\mathbb{R}^N_>$ defined by a polynomial with $n+k+1$ monomials whose exponent vectors have n-dimensional affine span has fewer than*

$$\frac{e^2+3}{4} \cdot 2^{\binom{k+1}{2}} 2^n n^{k+1}$$

connected components.

Both our method and that of Perucci estimate the number of compact components in the intersection of the hypersurface with linear spaces which are chosen to preserve the sparse structure so that the fewnomial bound still applies. For this same reason, this same method was used by Benedetti, Loeser, and Risler [1] to bound the number of components of a real algebraic set in terms of the degrees of its defining polynomials.

Observe that if the exponent vectors of the monomials in a polynomial f in N variables have n-dimensional affine span, then there is a monomial change of variables on $\mathbb{R}^N_>$ so that f becomes a polynomial in the first n variables, and thus our hypersurface becomes a cylinder

$$\mathbb{R}^{N-n}_< \times \{x \in \mathbb{R}^n_> \mid f(x) = 0\},$$

which shows that it suffices to prove Theorem 2 when $n = N$. That is, for smooth hypersurfacs in $\mathbb{R}^n$ defined by polynomials with $n+k+1$ monomial terms whose exponent vectors affinely span $\mathbb{R}^n$.

Let $\kappa(n,k)$ be the maximum number of compact connected components of such a smooth hypersurface and let $\tau(n,k)$ be the maximal number of connected components of such a hypersurface. We deduce Theorem 2 from the following combinatorial lemma.

LEMMA 3. $\tau(n,k) \leq \sum_{i=0}^{n-1} \binom{n+1}{i} \kappa(n-i, k+1)$.

Proof of Theorem 2. Bihan and Sottile [3] proved that

$$\kappa(n,k) \leq \frac{e^2+3}{8} 2^{\binom{k}{2}} n^k .$$

Substituting this into Lemma 3 bounds $\tau(n,k)$ by

$$\frac{e^2+3}{8} 2^{\binom{k+1}{2}} \sum_{i=0}^{n-1} \binom{n+1}{i} (n-i)^{k+1}$$

$$< \frac{e^2+3}{8} 2^{\binom{k+1}{2}} n^{k+1} \sum_{i=0}^{n+1} \binom{n+1}{i} = \frac{e^2+3}{8} 2^{\binom{k+1}{2}} n^{k+1} 2^{n+1} . \quad \square$$

Proof of Lemma 3. Let f be a polynomial in the variables $x_1, \ldots, x_n$ which has $n+k+1$ distinct monomials whose exponent vectors affinely span $\mathbb{R}^n$ and suppose that $f(x) = 0$ defines a smooth hypersurface X in $\mathbb{R}^n_>$. We may apply a monomial change of coordinates to $\mathbb{R}^n_>$ and assume that $1, x_1, x_2, \ldots, x_n$ are among the monomials appearing in f.

Suppose that $\varepsilon := (\varepsilon_0, \varepsilon_1, \ldots, \varepsilon_n) \in \mathbb{R}^{1+n}_>$ with $\varepsilon_0\varepsilon_1 \cdots \varepsilon_n \neq 1$. Define hypersurfaces $H_0, H_1, \ldots, H_n$ of $\mathbb{R}^n_>$ by

$$H_0 := \{x \in \mathbb{R}^n_> \mid x_1 \cdots x_n \varepsilon_0 = 1\} \quad \text{and}$$
$$H_i := \{x \mid x_i = \varepsilon_i\} \quad \text{for} \quad i = 1, \ldots, n .$$

The transformation Log: $\mathbb{R}^n_> \to \mathbb{R}^n$ defined on each coordinate by $x_i \mapsto \log(x_i)$ sends the hypersurfaces $H_0, H_1, \ldots, H_n$ to hyperplanes in general position. That is, if $S \subset \{0, 1, \ldots, n\}$ and we define $H_S := \cap_{i \in S} H_i$, then $\text{Log}(H_S)$ is an affine linear space of dimension $n-|S|$. Moreover, the complement of the union of hypersurfaces H_i has $2^{n+1} - 1$ connected components, exactly one of which is bounded.

If we restrict f to some H_S, we will obtain a new polynomial f_S in $n - |S|$ variables with at most $1+(n-|S|)+(k+1)$ monomials. Indeed, if $i \in S$ with $i \neq 0$, then the equation $x_i = \varepsilon_i$ allows us to eliminate both the variable and the monomial x_i from f. If however $0 \in S$, then we pick an index $j \notin S$ and use $x_1 \cdots x_n \varepsilon_0 = 1$ to eliminate the variable x_j, which will not necessarily eliminate a monomial from f. For almost all ε, the polynomial f_S defines a smooth hypersurface X_S of H_S.

We may choose ε small enough so that every compact connected component of X lies in the bounded region of the complement of the hypersurfaces H_i, and every noncompact connected component of X meets some hypersurface H_i. Shrinking ε if necessary, we can ensure that every bounded component of X_S lies in the bounded region of the complement of $H_j \cap H_S$ for $j \notin S$, and every unbounded component meets some $H_j \cap H_S$ for $j \notin S$.

Given a connected component C of X, the subsets $S \subset \{0,1,\ldots,n\}$ such that C meets H_S form a simplicial complex. If S represents a maximal simplex in this complex, then $C \cap H_S$ is a union of compact components of X_S, and $|S| < n$ as H_S is not a point. Thus the number of connected components of X is bounded by the sum of the numbers of compact components of X_S for all $S \subset \{0,1,\ldots,n\}$ with $n > |S|$. Since each f_S has at most $1+(n-|S|)+(k+1)$ monomials, this sum is bounded by the sum in the statement of the lemma. □

REMARK 1.2. If f contains a monomial $x^a := x_1^{a_1}x_2^{a_2}\cdots x_n^{a_n}$ with no $a_i = 0$, then we can alter the proof of Lemma 3 to obtain a bound of

$$\frac{e^2+3}{4}\cdot 2^{\binom{k}{2}}2^n n^k$$

connected components for the hypersurface defined by f.

The basic idea is that if we redefine H_0 to be

$$H_0 \;:=\; \{x \in \mathbb{R}^n_> \mid x^a\varepsilon_0 = 1\}\,.$$

then the polynomials f_S on H_S have only $1+(n-|S|)+k$ monomials, and so we estimate the number of compact components of X_S by $\kappa(n-i,k)$ instead of $\kappa(n-i,k+1)$.

Acknowledgment. We thank Alicia Dickenstein and Daniel Perucci whose comments inspired us to find the correct statement and proof of Theorem 2.

REFERENCES

[1] R. BENEDETTI, F. LOESER, AND J.-J. RISLER, *Bounding the number of connected components of a real algebraic set*, Discrete Comput. Geom., **6** (1991), pp. 191–209.

[2] F. BIHAN, *Polynomial systems supported on circuits and dessins d'enfants*, 2005. Journal of the London Mathematical Society, to appear.

[3] F. BIHAN AND F. SOTTILE, *New fewnomial upper bounds from Gale dual polynomial systems*, 2006. Moscow Mathematical Journal, to appear. math.AG/0609544.

[4] G.-M. GREUEL, G. PFISTER, AND H. SCHÖNEMANN, SINGULAR *3.0*, A Computer Algebra System for Polynomial Computations, Centre for Computer Algebra, University of Kaiserslautern, 2005. http://www.singular.uni-kl.de.

[5] A. KHOVANSKII, *A class of systems of transcendental equations*, Dokl. Akad. Nauk. SSSR, **255** (1980), pp. 804–807.

[6] ———, *Fewnomials*, Trans. of Math. Monographs, 88, AMS, 1991.

[7] T.-Y. LI, J.M. ROJAS, AND X. WANG, *Counting real connected components of trinomial curve intersections and m-nomial hypersurfaces*, Discrete Comput. Geom., **30** (2003), pp. 379–414.

[8] D. NAPOLETANI, *A power function approach to Kouchnirenko's conjecture*, in Symbolic computation: solving equations in algebra, geometry, and engineering (South Hadley, MA, 2000), Vol. **286** of Contemp. Math., Amer. Math. Soc., Providence, RI, 2001, pp. 99–106.

[9] D. PERRUCCI, *Some bounds for the number of components of real zero sets of sparse polynomials*, Discrete Comput. Geom., **34** (2005), pp. 475–495.

[10] W. POHL AND M. WENK, solve.lib, 2006. A SINGULAR 3.0 library for Complex Solving of Polynomial Systems.

INTERSECTIONS OF SCHUBERT VARIETIES AND OTHER PERMUTATION ARRAY SCHEMES

SARA BILLEY* AND RAVI VAKIL†

Abstract. Using a blend of combinatorics and geometry, we give an algorithm for algebraically finding all flags in any zero-dimensional intersection of Schubert varieties with respect to three transverse flags, and more generally, any number of flags. The number of flags in a triple intersection is also a structure constant for the cohomology ring of the flag manifold. Our algorithm is based on solving a limited number of determinantal equations for each intersection (far fewer than the naive approach in the case of triple intersections). These equations may be used to compute Galois and monodromy groups of intersections of Schubert varieties. We are able to limit the number of equations by using the *permutation arrays* of Eriksson and Linusson, and their permutation array varieties, introduced as generalizations of Schubert varieties. We show that there exists a unique permutation array corresponding to each realizable Schubert problem and give a simple recurrence to compute the corresponding rank table, giving in particular a simple criterion for a Littlewood-Richardson coefficient to be 0. We describe pathologies of Eriksson and Linusson's permutation array varieties (failure of existence, irreducibility, equidimensionality, and reducedness of equations), and define the more natural *permutation array schemes*. In particular, we give several counterexamples to the Realizability Conjecture based on classical projective geometry. Finally, we give examples where Galois/monodromy groups experimentally appear to be smaller than expected.

Key words. Schubert varieties, permutation arrays, Littlewood-Richardson coefficients.

AMS(MOS) subject classifications. Primary 14M17; Secondary 14M15.

1. Introduction. A typical *Schubert problem* asks how many lines in three-space meet four generally chosen lines. The answer, two, may be obtained by computation in the cohomology ring of the Grassmannian variety of two-dimensional planes in four-space. Such questions were considered by H. Schubert in the nineteenth century. During the past century, the study of the Grassmannian has been generalized to the flag manifold where one can ask analogous questions.

The flag manifold $\mathcal{F}l_n(K)$ parameterizes the complete flags

$$F_\bullet = \{\{0\} = F_0 \subset F_1 \subset \cdots \subset F_n = K^n\}$$

where F_i is a vector space of dimension i. (*Unless otherwise noted, we will work over an arbitrary base field* K. The reader, and Schubert, is welcome to assume $K = \mathbb{C}$ throughout. For a general field, we should use the Chow ring rather than the cohomology ring, but they agree for $K = \mathbb{C}$. For simplicity, we will use the term "cohomology" throughout.)

* Department of Mathematics, University of Washington, Seattle, WA 98195 (billey@math.washington.edu); supported by NSF grant DMS-9983797.

† Department of Mathematics, Stanford University, Stanford, CA 94305 (vakil@math.stanford.edu); supported by NSF grant DMS-0238532.

A modern Schubert problem asks how many flags have relative position u, v, w with respect to three generally chosen fixed flags $X_\bullet$, $Y_\bullet$ and $Z_\bullet$. One concrete solution to this problem, due to Lascoux and Schützenberger [Lascoux and Schützenberger, 1982], is to compute a product of Schubert polynomials and expand in the Schubert polynomial basis. The coefficient indexed by u, v, w is the solution. This corresponds to a computation in the cohomology ring of the flag variety. (Caution: this solution is known to work only in characteristic 0, due to the potential failure of the Kleiman-Bertini theorem in positive characteristic, cf. [Vakil, 2006b, Sect. 2].) The quest for a combinatorial rule for expanding these products is a long-standing open problem, and corresponds to the multiplication rule for Schubert polynomials.

The main goal of this paper is to describe a method for directly identifying all flags in $X_u(F_\bullet) \cap X_v(G_\bullet) \cap X_w(H_\bullet)$ when $\ell(u)+\ell(v)+\ell(w) = \binom{n}{2}$, thereby computing $c_{u,v,w}$ explicitly. This method extends to Schubert problems with more than three flags, and more generally to parameter spaces of flags in given relative position. The only geometrically reasonable meaning of "given relative position" is the specification of a "rank table" of intersection dimensions, tracking how the pieces of the various flags meet. Achievable or realizable rank tables yield unique "permutation arrays", and indeed this problem motivated their definition by Eriksson and Linusson. These permutation arrays are closely related to the checker boards used in [Vakil, 2006a, Vakil, 2006b]. The resulting permutation array varieties are natural generalizations of Schubert varieties to an arbitrary number of flags. The advantages of our method are further described in Remark 5.1.

The benefit of permutation arrays is that the elements identify the minimal jumps in dimension, and therefore naturally correspond to critical vectors in the geometry. We use the data from the permutation array to identify and solve a collection of determinantal equations for the permutation array varieties, allowing us to solve Schubert problems explicitly and effectively, for example allowing us to compute Galois/monodromy groups. Maple code for solving Schubert problems using permutation arrays is available at

`http://www.math.washington.edu/~billey/papers/maple.code/`

We show that permutation array varieties may be badly behaved. For example, their equations are not always reduced or irreducible, so we argue that the "correct" generalization of Schubert varieties are permutation array *schemes*. We describe pathologies of these varieties/schemes, and show that they are not irreducible nor even equidimensional in general, making a generalization of the Bruhat order problematic. We also give counterexamples to Eriksson and Linusson's Realizability Conjecture 4.1.

We emphasize that the pathologies described here are not an artifact of permutation arrays; permutation arrays are equivalent to tables of intersection dimensions. Permutation arrays are much more manageable as data sets than the full table of intersection dimensions.

On one hand our results are bad news for permutation arrays: the hope that they would predict which rank tables (tables of all intersection dimensions) are possible does not hold true, and this deep question remains open. This difficulty of this problem is very similar to the problem of determining which matroids are realizable. On the other hand, by highlighting key linear-algebraic data, they provide more geometric information about a Schubert problem which can be used for computation. In many of the examples we have tried, the new approach is more effective than any earlier naive approach, sometimes requiring no calculations at all beyond construction of the permutation array. It is an interesting open problem to determine which method is most effective for large Schubert problems. Furthermore, permutation arrays are a "complete flag analog" of Vakil's checkerboards [Vakil, 2006a]. So, one could ask if there exists a rule for multiplying Schubert classes based on these arrays.

Varieties based on rank tables have appeared in several other places in the literature as well, including [Eisenbud and Saltman, 1987] [Fulton, 1991, Magyar, 2005, Magyar and van der Kallen, 1999].

The outline of the paper is as follows. In Section 2, we review Schubert varieties and the flag manifold. In Section 3, we review the construction of permutation arrays and the Eriksson-Linusson algorithm for generating all such arrays. In Section 4, we describe permutation varieties and their pathologies, and explain why their correct definition is as schemes. In Section 5, we describe how to use permutation arrays to solve Schubert problems and give equations for certain intersections of Schubert varieties. In Section 6, we give two examples of an algorithm for computing triple intersections of Schubert varieties and thereby computing the cup product in the cohomology ring of the flag manifold. The equations we give also allow us to compute Galois and monodromy groups for intersections of Schubert varieties; we describe this application in Section 7. To our knowledge, this is the first use of the Hilbert irreducibility theorem to compute monodromy groups. Our computations lead to examples where the Galois/monodromy group is "smaller than expected".

2. The flag manifold and Schubert varieties. In this section we briefly review the notation and basic concepts for flag manifolds and Schubert varieties. We refer the reader to one of the following books for further background information: [Fulton, 1997, Macdonald, 1991, Manivel, 1998, Gonciulea and V. Lakshmibai, 2001, Kumar, 2002].

As described earlier, the flag manifold $\mathcal{F}l_n = \mathcal{F}l_n(K)$ parametrizes the complete flags

$$F_\bullet = \{\{0\} = F_0 \subset F_1 \subset \cdots \subset F_n = K^n\}$$

where F_i is a vector space of dimension i over the field K. $\mathcal{F}l_n$ is a smooth projective variety of dimension $\binom{n}{2}$. A complete flag is determined by an ordered basis $(f_1, \ldots, f_n)$ for K^n by taking $F_i = \text{span}(f_1, \ldots, f_i)$.

Two flags $[F_\bullet], [G_\bullet] \in \mathcal{F}l_n$ are *in relative position* $w \in S_n$ when

$$\dim(F_i \cap G_j) = \operatorname{rank} w[i,j] \qquad \text{for all } 1 \leq i,j \leq n$$

where $w[i,j]$ is the principal submatrix of the permutation matrix for w with lower right hand corner in position (i,j). We use the notation

$$\operatorname{pos}(F_\bullet, G_\bullet) = w.$$

Warning: in order to use the typical meaning for a principal submatrix we are using a nonstandard labeling of a permutation matrix. The permutation matrix we associate to w has a 1 in the $w(i)$th row of column $n-i+1$ for $1 \leq i \leq n$. For example, the matrix associated to $w = (5,3,1,2,4)$ is

$$\begin{bmatrix} 0 & 0 & 1 & 0 & 0 \\ 0 & 1 & 0 & 0 & 0 \\ 0 & 0 & 0 & 1 & 0 \\ 1 & 0 & 0 & 0 & 0 \\ 0 & 0 & 0 & 0 & 1 \end{bmatrix}.$$

If $\operatorname{pos}(F_\bullet, G_\bullet) = (5,3,1,2,4)$ then $\dim(F_2 \cap G_3) = 2$ and $\dim(F_3 \cap G_2) = 1$.

Define a *Schubert cell* with respect to a fixed flag $F_\bullet$ in $\mathcal{F}l_n$ to be

$$\begin{aligned} X^\circ_w(F_\bullet) &= \{G_\bullet \mid F_\bullet \text{ and } G_\bullet \text{ have relative position } w\} \\ &= \{G_\bullet \mid \dim(F_i \cap G_j) = \operatorname{rk} w[i,j]\}. \end{aligned}$$

Using our labeling of a permutation matrix, the codimension of X°_w is equal to the length of w (the number of inversions in w), denoted $\ell(w)$. In fact, X°_w is isomorphic to the affine space $K^{\binom{n}{2}-\ell(w)}$. We say the flags $F_\bullet$ and $G_\bullet$ are in *transverse position* if $G_\bullet \in X_{\mathrm{id}}(F_\bullet)$. A randomly chosen flag will be transverse to any fixed flag $F_\bullet$ with probability 1 (using any reasonable measure, assuming the field is infinite).

The *Schubert variety* $X_w(F_\bullet)$ is the closure of $X^\circ_w(F_\bullet)$ in $\mathcal{F}l_n$. Schubert varieties may also be written in terms of rank conditions:

$$X_w(F_\bullet) = \{G_\bullet \mid \dim(F_i \cap G_j) \geq \operatorname{rk} w[i,j]\}. \tag{2.1}$$

If the flags $F_\bullet$ and $G_\bullet$ are determined by ordered bases for K^n then these inequalities can be rephrased as determinantal equations on the coefficients in the bases [Fulton, 1997, 10.5, Ex. 10, 11]. Of course this allows one in theory to solve all Schubert problems, but the number and complexity of the equations conditions grows quickly to make this prohibitive for large n or d. See Section 5.2 for more details.

We remark that the rank equations in (2.1) are typically written in terms of an increasing rank function in the literature as we have done. However, when one wants to write down polynomial equations which vanish on this set, one must use a decreasing rank function. A rank function strictly less than k on a matrix means that every $k \times k$ determinantal minor vanishes, while a rank function strictly greater than k means that SOME $j \times j$ minor for $j \geq k$ does NOT vanish. The first description defines a closed subvariety, but the second condition does not. Luckily the rank functions that we are interested in are the coranks of the matrices with the ordered basis reversed so when we need to explicitly present polynomial equations that define a Schubert variety, we will use decreasing rank functions.

The cohomology ring $H^*(\mathcal{F}l_n)$ of $\mathcal{F}l_n$ is isomorphic to

$$\mathbb{Z}[x_1, \ldots, x_n]/\langle e_1, e_2, \ldots, e_n \rangle$$

where e_i is the ith elementary symmetric function on $x_1, \ldots, x_n$. For details see [Fulton, 1997, 10.2, B.3]. The cycles $[X_u]$ corresponding to Schubert varieties form a $\mathbb{Z}$-basis for the ring. The class $[X_u] := [X_u(F_\bullet)]$ is independent of the choice of base flag. The product is defined by

$$[X_u] \cdot [X_v] = [X_u(F_\bullet) \cap X_v(G_\bullet)]$$

where $F_\bullet$ and $G_\bullet$ are in transverse position. Speaking informally, $X_u(F_\bullet) \cap X_v(G_\bullet)$ is a union of irreducible components which are GL_n-translates of Schubert varieties. Therefore, in the cohomology ring, the expansion of a product of Schubert cycles

$$[X_u] \cdot [X_v] = \sum_{\ell(w) = \ell(u) + \ell(v)} c_{u,v}^w [X_w] \tag{2.2}$$

has nonnegative integer coefficients in the basis of Schubert cycles.

A simpler geometric interpretation of the coefficients $c_{u,v}^w$ may be given in terms of triple intersections [Fulton, 1997, 10.2]. There exists a perfect pairing on $H^*(\mathcal{F}l_n)$ such that

$$[X_w] \cdot [X_y] = \begin{cases} [X_{w_\circ}] & y = w_\circ w \\ 0 & y \neq w_\circ w, \ \ell(y) = \binom{n}{2} - \ell(w). \end{cases} \tag{2.3}$$

Here $w_\circ = (n, n-1, \ldots, 1)$ is the longest permutation in S_n, of length $\binom{n}{2} = \dim(\mathcal{F}l_n)$, and $[X_{w_\circ}]$ is the class of a point. Combining equations (2.2) and (2.3) we have

$$[X_u] \cdot [X_v] \cdot [X_{w_\circ w}] = c_{u,v}^w [X_{w_\circ}].$$

In characteristic 0, $c_{u,v}^w$ counts the number of points $[E_\bullet] \in \mathcal{F}l_n$ in the variety

$$X_u(F_\bullet) \cap X_v(G_\bullet) \cap X_{w_\circ w}(H_\bullet) \tag{2.4}$$

when $\ell(u) + \ell(v) + \ell(w_\circ w) = \binom{n}{2}$ and $F_\bullet, G_\bullet, H_\bullet$ are three generally chosen flags. Note, it is not sufficient to assume the three flags are pairwise transverse in order to get the expected number of points in the intersection. There can be additional dependencies among the subspaces of the form $F_i \cap G_j \cap H_k$.

The main goal of this article is to describe a method to find all flags in a general d-fold intersection of Schubert varieties when the intersection is zero-dimensional. Enumerating the flags found explicitly in a triple intersection would give the numbers $c_{u,v}^{w}$. We will use the permutation arrays defined in the next section to identify a different set of equations defining the intersections of Schubert varieties which are easier to solve.

3. Permutation arrays. In [Eriksson and Linusson, 2000a] and [Eriksson and Linusson, 2000b], Eriksson and Linusson develop a d dimensional analog of a permutation matrix. One way to generalize permutation matrices is to consider all d-dimensional arrays of 0's and 1's with a single 1 in each hyperplane with a single fixed coordinate. They claim that a better way is to consider a permutation matrix to be a two-dimensional array of 0's and 1's such that the rank of any principal minor is equal to the number of occupied rows in that submatrix or equivalently equal to the number of occupied columns in that submatrix. The locations of the 1's in a permutation matrix will be the elements in the corresponding permutation array. We will summarize their work here and refer the reader to their well-written paper for further details.

Let P be any collection of points in $[n]^d := \{1, 2, \ldots, n\}^d$. We will think of these points as the locations of dots in an $[n]^d$-*dot array*. Define a partial order on $[n]^d$ by

$$x = (x_1, \ldots, x_d) \preceq y = (y_1, \ldots, y_d),$$

read "x is *dominated* by y", if $x_i \leq y_i$ for all $1 \leq i \leq d$. This poset is a lattice with meet and join operation defined by

$$\begin{aligned} x \vee y = z \qquad & \text{if } z_i = \max(x_i, y_i) \text{ for all } i \\ x \wedge y = z \qquad & \text{if } z_i = \min(x_i, y_i) \text{ for all } i. \end{aligned}$$

These operations extend to any set of points R by taking $\bigvee R = z$ where z_i is the the maximum value in coordinate i over the whole set, and similarly for $\bigwedge R$.

Let $P[y] = \{x \in P \mid x \preceq y\}$ be the *principal subarray* of P containing all points of P which are dominated by y. Define

$$\mathrm{rk}_j P = \#\{1 \leq k \leq n \mid \text{there exists } x \in P \text{ with } x_j = k\}.$$

P is *rankable* of *rank* r if $\mathrm{rk}_j P = r$ for all $1 \leq j \leq d$. P is *totally rankable* if every principal subarray of P is rankable.

For example, with $n = 4$, $d = 3$ the following example is a totally rankable dot array:

$$\{(3,4,1), (4,2,2), (1,4,3), (3,3,3), (2,3,4), (3,2,4), (4,1,4)\}.$$

We picture this as four 2-dimensional slices, where the first one is "slice 1" and the last is "slice 4":

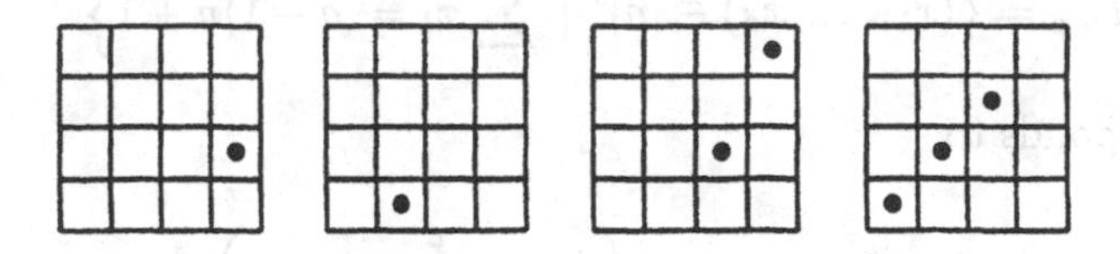

Thus $(3,4,1)$ corresponds to the dot in the first slice on the left.

The array $\{(3,4,1),(4,2,2),(1,4,3)\}$ is not rankable since it has only two distinct values appearing in the second index and three in the first and third.

Many pairs P, P' of totally rankable dot arrays are *rank equivalent*, i.e. $\text{rk}_j P[x] = \text{rk}_j P'[x]$, for all x and j. However, among all rank equivalent dot arrays there is a unique one with a minimal number of dots [Eriksson and Linusson, 2000a, Prop. 4.1]. In order to characterize the minimal totally rankable dot arrays, we give the following two definitions. We say a position x is *redundant* in P if there exists a collection of points $R \subset P$ such that $x = \bigvee R$, $\#R > 1$, and every $y \in R$ has at least one $y_i = x_i$. We say a position x is *covered* by dots in P if x is redundant for some $R \subset P$, $x \notin R$, and for each $1 \leq j \leq d$ there exists some $y \in R$ such that $y_j < x_j$. We show in Lemma 3.1 that it suffices to check only subsets R of size at most d when determining if a position is redundant or covered.

THEOREM 3.1. *[Eriksson and Linusson, 2000b, Theorem 3.2] Let P be a dot array. The following are equivalent:*

1. *P is totally rankable.*
2. *Every two dimensional projection of every principal subarray is totally rankable.*
3. *Every redundant position is covered by dots in P.*
4. *If there exist dots in P in positions y and z and integers i, j such that $y_i < z_i$ and $y_j = z_j$, then there exists a dot in some position $x \preceq (y \vee z)$ such that $x_i = z_i$ and $x_j < z_j$.*

Define a *permutation array* in $[n]^d$ to be a totally rankable dot array of rank n with no redundant dots (or equivalently, no covered dots). The permutation arrays are the unique representatives of each rank equivalence class of totally rankable dot arrays with no redundant dots. These arrays are Eriksson and Linusson's analogs of permutation matrices.

The definition of permutation arrays was motivated because they include the possible relative configurations of flags:

THEOREM 3.2. *[Eriksson and Linusson, 2000b, Thm. 3.1] Given flags $E^1_\bullet, E^2_\bullet, \ldots, E^d_\bullet$, there exists an $[n]^d$-permutation array P describing the rank table of all intersection dimensions as follows. For each $x \in [n]^d$,*

$$\text{rk}(P[x]) = \dim\left(E^1_{x_1} \cap E^2_{x_2} \cap \cdots \cap E^d_{x_d}\right). \tag{3.1}$$

A special case is the permutation array corresponding to n generally chosen flags, which we denote the *transverse permutation array*

$$T_{n,d} = \left\{(x_1, \ldots, x_d) \in [n]^d \mid \sum x_i = (d-1)n + 1\right\}. \tag{3.2}$$

This corresponds to

$$\mathrm{rk}(T_{n,d}[x]) = \max\left(0, n - \sum_{i=1}^{d}(n - x_i)\right).$$

Eriksson and Linusson give an algorithm for producing all permutation arrays in $[n]^d$ recursively from the permutation arrays in $[n]^{d-1}$ [Eriksson and Linusson, 2000b, Sect. 2.3]. We review their algorithm, which we call the *EL-algorithm* below, as this is key to our algorithm for intersecting Schubert varieties.

Let A be any antichain of dots in P under the dominance order. Let $C(A)$ be the set of positions covered by dots in A. Define the *downsizing* operator $D(A, P)$ with respect to A on P to be the result of the following process.

1. Set $Q_1 = P \setminus A$.
2. Set $Q_2 = Q_1 \cup C(A)$.
3. Set $D(A, P) = Q_2 \setminus R(Q_2)$ where $R(Q)$ is the set of redundant positions of Q.

The downsizing of a totally rankable array P is *successful* if the resulting array is again totally rankable and has rank $\mathrm{rk}(P) - 1$.

THEOREM 3.3. ***(The EL-Algorithm)*** *For each pair n, d of positive integers, the set of all permutation arrays in $[n]^d$ can be obtained by the following algorithm:*

1. For each permutation array P_n in $[n]^{d-1}$.

2. If P_n, ..., P_i have been defined, and $i > 1$, then choose an antichain A_i of dots in P_i such that the downsizing $D(A_i, P_i)$ is successful. Set $P_{i-1} = D(A_i, P_i)$.

3. Set $A_1 = P_1$.

4. Set $P = \{(x_1, \ldots, x_{d-1}, i) \mid (x_1, \ldots, x_{d-1}) \in A_i\}$. Add P to the list of permutation arrays in $[n]^d$ constructed thus far.

Furthermore, each permutation array P is constructed from a unique P_n in $[n]^{d-1}$ and a unique sequence of anti-chains.

For example, starting with the 2-dimensional array

$$\{(1,4), (2,3), (3,1), (4,2)\}$$

corresponding to the permutation $w = (1, 2, 4, 3)$, we run through the algorithm as follows. (In the figure, dots correspond to elements in P and circled dots correspond to elements in A.)

$P_4 = \{(1,4),(2,3),(3,1),(4,2)\}$ $\qquad$ $A_4 = \{(1,4),(2,3)\}$
$P_3 = \{(2,4),(3,1),(4,2)\}$ $\qquad$ $A_3 = \{(3,1)\}$
$P_2 = \{(2,4),(4,2)\}$ $\qquad$ $A_2 = \{(2,4),(4,2)\}$
$P_1 = \{(4,4)\}$ $\qquad$ $A_1 = \{(4,4)\}$

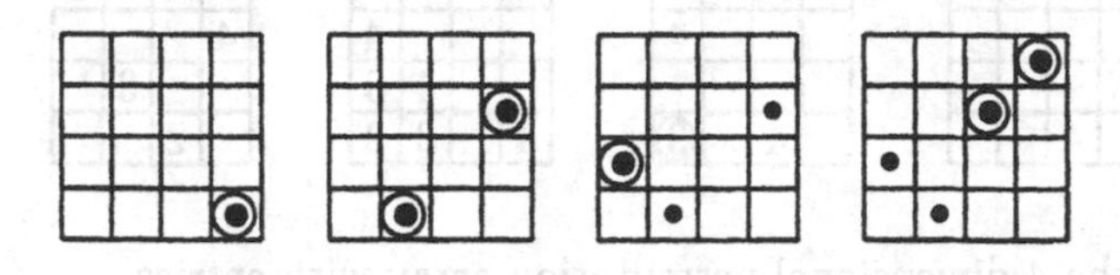

This produces the 3-dimensional array

$$P = \{(4,4,1),(2,4,2),(4,2,2),(3,1,3),(1,4,4),(2,3,4)\}.$$

We prefer to display 3-dimensional dot-arrays as 2-dimensional number arrays as in [Eriksson and Linusson, 2000b, Vakil, 2006a] where a square (i,j) contains the number k if $(i,j,k) \in P$. The previous example is represented by

			4
		4	2
3			
	2		1

.

Note that there is at most one number in any square if the number-array represents a permutation array: by Theorem 3.1 Part 4, if two dots y, z in a totally rankable array P existed such that $y_1 = z_1, y_2 = z_2, y_3 < z_3$, then there exists a third dot $x \preceq (y \vee z) = z$ in P with $x_3 = z_3$ and $x_i < y_i$ for $i = 1$ or 2, but this implies that z is redundant for the set $R = \{x, y\}$, hence P is not a permutation array.

COROLLARY 3.1. *In Theorem 3.3, each P_i is an $[n]^{d-1}$-permutation array of rank i. Furthermore, if P determines the rank table for flags $E^1_\bullet, \ldots, E^d_\bullet$, then P_i determines the rank table for $E^1_\bullet, \ldots, E^{d-1}_\bullet$ intersecting the vector space E^d_i, i.e.*

$$\mathrm{rk}\,(P_i[x]) = \dim\left(E^1_{x_1} \cap E^2_{x_2} \cap \cdots \cap E^{d-1}_{x_{d-1}} \cap E^d_i\right).$$

Proof. P_i is the permutation array obtained from the projection

$$\{(x_1, \ldots, x_d) \mid (x_1, \ldots, x_d, x_{d+1}) \in P \text{ and } x_{d+1} \leq i\}$$

by removing all repeated or covered elements. □

To represent a 4-dimensional permutation array, we often draw the n 3-dimensional permutation arrays $P_1, \ldots, P_n$ from the EL-algorithm. For example,

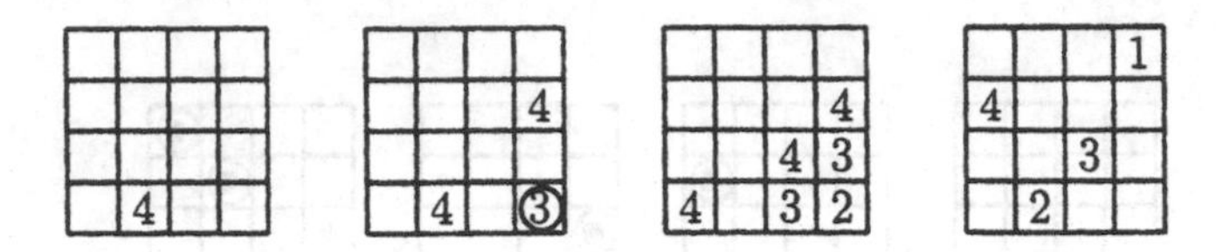

represents the 4-dimensional permutation array with entries

$$(4,2,4,1),(2,4,4,2),(4,4,3,2),(3,3,4,3),(3,4,3,3),(4,3,3,3),$$
$$(4,4,2,3),(1,4,1,4),(2,1,4,4),(3,3,3,4),(4,2,2,4).$$

We finish this section with a substantial improvement on the speed to the Eriksson-Linusson algorithm. In Step 2 of Theorem 3.3, one must find all positions covered by a subset of points in the antichain A_i. This appears to require on the order of $2^{|A_i|}$ computations. However, here we show that subsets of size at most d are sufficient.

LEMMA 3.1. *A position $x \in [n]^d$ is covered (or equivalently, redundant) in a permutation array P if and only if there exists a subset S with $|S| \leq d$ which covers x.*

Proof. Assume x is covered by a set $Y = \{y^1, y^2, \ldots, y^k\}$ for $k > d$. That is,

- For each position $1 \leq j \leq d$, there exists a y^i such that $y^i_j < x_j$ and there exists a y^l such that $y^l_j = x_j$.
- For each $y^i \in Y$, there exists a j such that $y^i_j < x_j$ and there exists an l such that $y^i_l = x_l$.

Consider a complete bipartite graph with left vertices labeled by Y and right vertices labeled by $\{x_1, \ldots, x_d\}$. Color the edge from y^i to x_j red if $y^i_j = x_j$, and blue if $y^i_j < x_j$. Since $x = \bigvee Y$, $y^i_j > x_j$ is not possible. This is a complete bipartite graph such that each vertex meets at least one red and one blue edge, and conversely any such complete bipartite graph with left vertices chosen from P and right vertices $\{x_1, \ldots, x_d\}$ corresponds to a covering of x.

We can easily bound the minimum size of a covering set for x to be at most $d+1$ as follows. Choose one red and one blue edge adjacent to x_1. Let S be the left end-points of these two edges. Vertex x_2 is connected to both elements of S in the complete bipartite graph. If the edges connecting x_2 to S are different colors, proceed to x_3. If the edges agree in color, choose one additional edge of a different color adjacent to x_2. Add its left endpoint to S. Continuing in this way for $x_3, \ldots, x_d$, we have $|S| \leq d+1$ and that x is covered by S.

Given a covering set S of size $d+1$, we now find a subset of size d which covers x. Say $x_{i_1}, x_{i_2}, \ldots, x_{i_k}$ are all the right vertices which are adjacent to a unique edge of either color. Let T be the left endpoints of all of these edges; these are necessary in any covering subset. Choose one vertex in $Y \setminus T$, say $\tilde{y}$. Each remaining x_j has at least two edges of each color, so we can choose one of each color which is not adjacent to $\tilde{y}$. The induced subgraph on $(S \setminus \{\tilde{y}\}, \{x_1, \ldots, x_d\})$ is again a complete bipartite graph where every vertex is adjacent to at least one red and one blue edge, hence $S \setminus \{\tilde{y}\}$ covers x. □

4. Permutation array varieties/schemes and their pathologies. In analogy with Schubert cells, for any $[n]^d$-permutation array P, Eriksson and Linusson define the *permutation array variety* X_P° to be the subset of $\mathcal{F}l_n^d = \{(E_\bullet^1, \ldots, E_\bullet^d)\}$ in "relative position P" [Eriksson and Linusson, 2000b, §1.2.2]. We will soon see why X_P° is a locally closed subvariety of $\mathcal{F}l_n^d$; this will reinforce the idea that the correct notion is of a permutation array *scheme*. The closure of X_P° will be defined by the rank equations

$$\operatorname{rk}(P[x]) \geq \dim\left(E_{x_1}^1 \cap E_{x_2}^2 \cap \cdots \cap E_{x_d}^d\right). \tag{4.1}$$

These rank equations can then be interpreted as determinantal equations as we explain below. These varieties/schemes will give a convenient way to manage the equations of intersections of Schubert varieties.

Based on many examples, Eriksson and Linusson [Eriksson and Linusson, 2000b, Conj. 3.2] conjectured the following statement.

REALIZABILITY CONJECTURE 4.1. *Every permutation array can be realized by flags. Equivalently, every X_P° is nonempty.*

This question is motivated by more than curiosity. A fundamental question is: *what are the possible relative configurations of d flags?* In other words: *what rank (intersection dimension) tables are possible?* For $d = 2$, the answer leads to the theory of Schubert varieties. By Theorem 3.2, each achievable rank table yields a permutation array, and the permutation arrays may be enumerated by Theorem 3.3. The Realizability Conjecture then says that we have fully answered this fundamental question. Failure of realizability would imply that we still have a poor understanding of how flags can meet.

The Realizability Conjecture is true for $d = 1, 2, 3$. For $d = 1$, the only permutation array variety is the flag variety. For $d = 2$, the permutation array varieties are the "generalized" Schubert cells (where the reference flag may vary). The case $d = 3$ follows from [Shapiro et al., 1997] (as described in [Eriksson and Linusson, 2000b, §3.2]), see also [Vakil, 2006a, §4.8]. The case $n \leq 2$ is fairly clear, involving only one-dimensional subspaces of a two-dimensional vector space (or projectively, points on $\mathbb{P}^1$), cf. [Eriksson and Linusson, 2000b, Lemma 4.3]. Nonetheless, the conjecture is false, and we give examples below which show the bounds $d \leq 3$ and $n \leq 2$ are maximal for such a realizability statement. We found it interesting that the combinatorics of permutation arrays prevent some naive attempts at counterexamples from working; somehow, permutation arrays see some subtle linear algebraic information, but not all.

Fiber permutation array varieties. If P is an $[n]^{d+1}$ permutation array, then there is a natural morphism $X_P^\circ \to \mathcal{F}l_n^d$ corresponding to "forgetting the last flag". We call the fiber over a point $(E_\bullet^1, \dots, E_\bullet^d)$ a *fiber permutation array variety*, and denote it $X_P^\circ(E_\bullet^1, \dots, E_\bullet^d)$. If the flags $E_\bullet^1, \dots, E_\bullet^d$ are chosen generally, we call the fiber permutation array variety a *generic fiber permutation array variety.* Note that a generic fiber permutation array variety is empty unless the projection of the permutation array to the "bottom hyperplane of P" is the transverse permutation array $T_{n,d}$, as this projection describes the relative positions of the first d flags.

The Schubert cells $X_w^\circ(E_\bullet^1)$ are fiber permutation array varieties, with $d = 2$. Also, any intersection of Schubert cells

$$X_{w_1}(E_\bullet^1) \cap X_{w_2}(E_\bullet^2) \cap \cdots \cap X_{w_d}(E_\bullet^d)$$

is a disjoint union of fiber permutation array varieties, and if the $E_\bullet^i$ are generally chosen, the intersection is a disjoint union of generic fiber permutation array varieties.

Permutation array varieties were introduced partially for this reason, to study intersections of Schubert varieties, and indeed that is the point of this paper. It was hoped that they would in general be tractable and well-behaved (cf. the Realizability Conjecture 4.1), but sadly this is not the case. The remainder of this section is devoted to their pathologies, and is independent of the rest of the paper.

Permutation array schemes. We first observe that the more natural algebro-geometric definition is of *permutation array schemes*: the set of d-tuples of flags in configuration P comes with a natural scheme structure, and it would be naive to expect that the resulting schemes are reduced. In other words, the "correct" definition of X_P° will contain infinitesimal information not present in the varieties. More precisely, the X_P° defined above may be defined scheme-theoretically by the equations (3.1), and these equations will *not* in general be all the equations cutting out the *set* X_P° (see the "Further Pathologies" discussion below). Those readers preferring to avoid the notion of schemes may ignore this definition. Other readers should re-define X_P° to be the scheme cut out by equations (3.1), which is a locally closed subscheme of $\mathcal{F}l_n^d$. More explicitly, (3.1) specifies certain rank conditions, which can be written in terms of equations as follows. Requiring that the rank of a matrix is r corresponds to requiring that all of the $(r+1) \times (r+1)$ minors vanish, and that some $r \times r$ minor does not vanish.

We now give a series of counterexamples to the Realizability Conjecture 4.1.

Counterexample 1. Eriksson and Linusson defined their permutation array varieties over $\mathbb{C}$, so we begin with a counterexample to realiz-

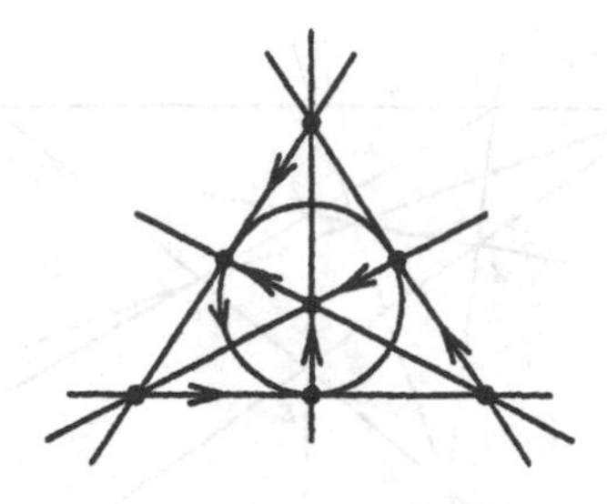

FIG. 1. *The Fano plane, and a bijection of points and lines (indicated by arrows from points to the corresponding line).*

ability over $K = \mathbb{C}$, and it may be read simply as an admonition to always consider a more general base field (or indeed to work over the integers). The Fano plane is the projective plane over the field $\mathbb{F}_2$, consisting of 7 lines $\ell_1, \ldots, \ell_7$ and 7 points $p_1, \ldots, p_7$. We may name them so that p_i lies on ℓ_i, as in Figure 1. Thus we have a configuration of 7 flags over $\mathbb{F}_2$. (This is a projective picture, so this configuration is in affine dimension $n = 3$, and the points p_i should be interpreted as one-dimensional subspaces, and the lines ℓ_j as two-dimensional subspaces, of K^3.) The proof of Theorem 3.2 is independent of the base field, so the rank table of intersection dimensions of the flags yields a permutation array. However, a classical and straightforward argument in projective coordinates shows that the configuration of Figure 1 may not be achieved over the complex numbers (or indeed over any field of characteristic not 2). In particular, this permutation array variety is not realizable over $\mathbb{C}$. In order to patch this counterexample, one might now restate the Realizability Conjecture 4.1 by saying that there always exists a field such that X_P° is nonempty. However, the problems have only just begun.

Counterexample 2. We next sketch an elementary counterexample for $n = 3$ and $d = 9$, over an arbitrary field, with the disadvantage that it requires a computer check. Recall Pappus' Theorem in classical geometry: if A, B, and C are collinear, and D, E, and F are collinear, and $X = AE \cap BD$, $Y = AF \cap CD$, and $Z = BF \cap CE$, then X, Y, and Z are collinear [Coxeter and Greitzer, 1967, §3.5]. The result holds over any field. A picture is shown in Figure 2. (Ignore the dashed arc and the stars for now.)

We construct an unrealizable permutation array as follows. We imagine that line YZ does not meet X. (In the figure, the starred line YZ "hops over" the point marked X.) We construct a counterexample with nine flags by letting the flags correspond to the nine lines of our "deformed Pappus configuration", choosing points on the lines arbitrarily. We then construct the rank table of this configuration, and verify that this corresponds to a valid permutation array. (This last step was done by computer.) This permutation array is not realizable, by Pappus' theorem.

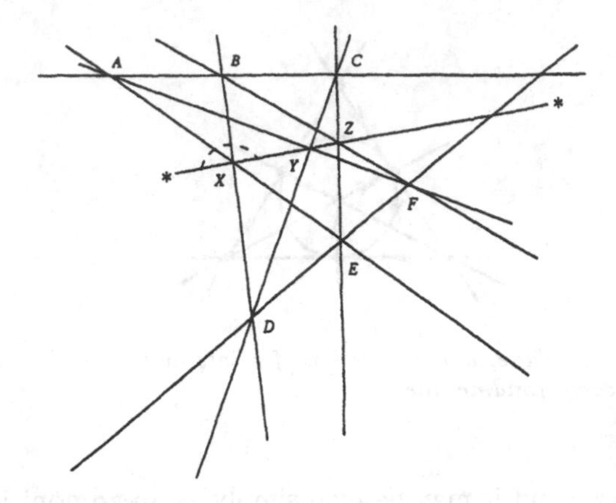

FIG. 2. *Pappus' Theorem, and a counterexample to Realizability in dimension* $d = 3$ *with* $n = 9$.

Counterexample 3. Our next example shows that realizability already fails for $n = 4$, $d = 4$. The projective intuition is as follows. Suppose ℓ_1, ℓ_2, ℓ_3, ℓ_4 are four lines in projective space, no three meeting in a point, such that we require ℓ_i and ℓ_j to meet, except (possibly) ℓ_3 and ℓ_4. This forces all 4 lines to be coplanar, so ℓ_3 and ℓ_4 *must* meet. Hence we construct an unrealizable configuration as follows: we "imagine" (as in Figure 3) that ℓ_3 and ℓ_4 don't meet. Again, we must turn the projective picture in $\mathbb{P}^3$ into linear algebra in 4-space, so the projective points in the figure correspond to one-dimensional subspaces, the projective lines in the figure correspond to two-dimensional subspaces of their respective flags, etc. Again, the tail of each arrow corresponds with the point which lies on the line the arrow follows. We construct the corresponding dot array:

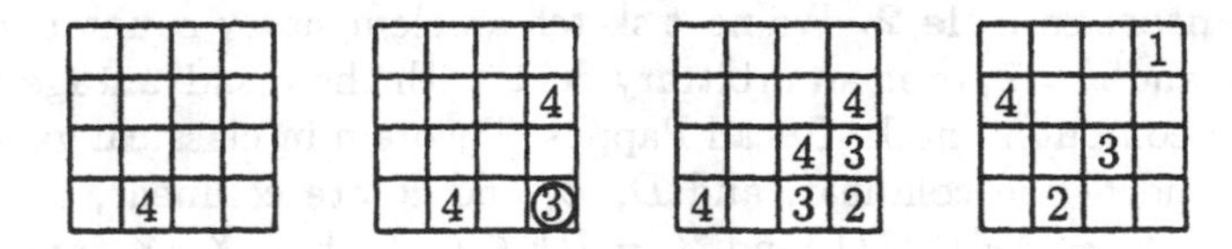

Here the rows represent the flag $F^1_\bullet$, columns represent the flag $F^2_\bullet$, numbers represent the flag $F^3_\bullet$, and the boards represent the flag $F^4_\bullet$. This is readily checked to be a permutation array. The easiest way is to compare it to the dot array for the "legitimate" configuration, where F^3_2 and F^4_2 *do* meet, and using the fact that this second array is a permutation array by Theorem 3.2. The only difference between the permutation array above and the "legitimate" one is that the circled 3 should be a 2.

Remark. Eriksson and Linusson have verified the Realizability Conjecture 4.1 for $n = 3$ and $d = 4$ [Eriksson and Linusson, 2000b, §3.1]. Hence the only four open cases left are $n = 3$ and $5 \leq d \leq 8$. These cases seem

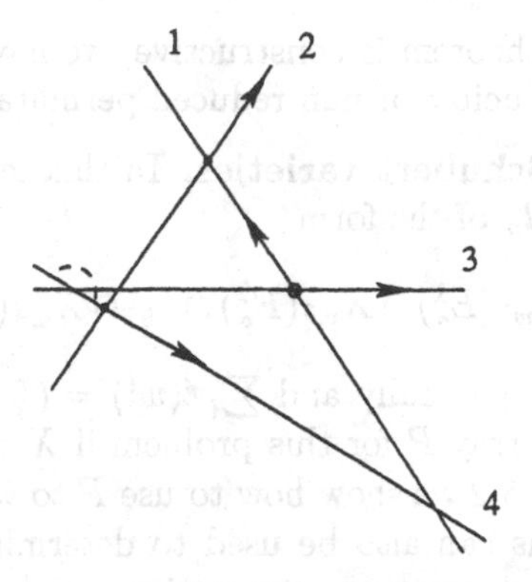

FIG. 3. *A counterexample to realizability with $n = d = 4$.*

simple, as they involve (projectively) between 5 and 8 lines in the plane. Can these remaining cases be settled?

Further pathologies from Mnëv's universality theorem: failure of irreducible and equidimensionality. Mnëv's universality theorem shows that permutation array schemes will be "arbitrarily" badly behaved in general, even for $n = 3$. Informally, Mnëv's theorem states that given any singularity type of finite type over the integers there is a configuration of projective lines in the plane such that the corresponding *permutation array scheme* has that singularity type. By a singularity type of finite type over the integers, we mean up to smooth parameters, any singularity cut out by polynomials with integer co-efficients in a finite number of variables. See [Mnëv, 1985, Mnëv, 1988] for the original sources, and [Vakil, 2006c, §3] for a precise statement and for an exposition of the version we need. (Mnëv's theorem is usually stated in a different language of course.)

In particular, (i) permutation array schemes need not be irreducible, answering a question raised in [Eriksson and Linusson, 2000b, §1.2.3]. They can have arbitrarily many components, indeed of arbitrarily many different dimensions. (ii) Permutation array schemes need not be reduced, i.e. they have genuine scheme-theoretic (or infinitesimal) structure not present in the variety. In other words, the definition of permutation array schemes is indeed different from that of permutation array varieties, and the equations (3.1) do not cut out the permutation array varieties scheme-theoretically. (iii) Permutation array schemes need not be equidimensional. Hence the hope that permutation array varieties/schemes might be well-behaved is misplaced. In particular, the notion of Bruhat order is problematic as already noted in [Eriksson and Linusson, 2000b]. We suspect, for example, that there exist two permutation array schemes X and Y such that Y is reducible, and some but not all components of Y lie in the closure of X.

Although Mnëv's theorem is constructive, we have not attempted to explicitly produce a reducible or non-reduced permutation array scheme.

5. Intersecting Schubert varieties. In this section, we consider a Schubert problem in $\mathcal{F}l_n$ of the form

$$X = X_{w^1}(E^1_\bullet) \cap X_{w^2}(E^2_\bullet) \cap \cdots \cap X_{w^d}(E^d_\bullet)$$

with $E^1_\bullet, \ldots, E^d_\bullet$ chosen generally and $\sum_i \ell(w^i) = \binom{n}{2}$. We show there is a unique permutation array P for this problem if X is nonempty, and we identify it. In Theorem 5.2 we show how to use P to write down equations for X. These equations can also be used to determine if $E^1_\bullet, \ldots, E^d_\bullet$ are sufficiently general for computing intersection numbers. The number of solutions will always be either infinite or no greater than the expected number. The expected number is achieved on a dense open subset of $\mathcal{F}l_n^d$. It may be useful for the reader to refer to the examples in Section 6 while reading this section.

THEOREM 5.1. *If X is 0-dimensional and nonempty, there exists a unique permutation array $P \subset [n]^{d+1}$ such that*

$$\dim\left(E^1_{x_1} \cap E^2_{x_2} \cap \cdots \cap E^d_{x_d} \cap F_{x_{d+1}}\right) = \mathrm{rk}P[x]$$

for all $F_\bullet \in X$ and all $x \in [n]^{d+1}$. Hence, X is equal to the fiber permutation array variety $X^\circ_P(E^1_\bullet, \ldots, E^d_\bullet)$.

It is natural to ask which permutation array this is, and this will be necessary for later computations. We describe the permutation array (in the guise of its rank table) in Subsection 5.1.

The generalization to the case where X has positive dimension is left to the interested reader; the permutation array then describes the generic behavior on every component of X. The argument below carries through essentially without change.

Proof. Consider the variety

$$X' := X_{w^1}(F_\bullet) \times_{\mathcal{F}l_n} X_{w^2}(F_\bullet) \times_{\mathcal{F}l_n} \cdots \times_{\mathcal{F}l_n} X_{w^d}(F_\bullet). \tag{5.1}$$

Here $F_\bullet$ is the flag parametrized by the base $\mathcal{F}l_n$. The "incidence variety" X' is a product of Schubert variety bundles over the flag variety, and hence clearly irreducible; its dimension is

$$\dim X' = (d+1)\binom{n}{2} - \sum_i \ell(w^i) = d\binom{n}{2}.$$

Let $E^i_\bullet$ be the flag parametrized by the ith factor of (5.1). To each point of X' there is a permutation array describing how the $d+1$ flags $E^1_\bullet, \ldots, E^d_\bullet, F_\bullet$ meet, i.e. with rank table

$$\dim\left(E^1_{x_1} \cap E^2_{x_2} \cap \cdots \cap E^d_{x_d} \cap F_{x_{d+1}}\right).$$

As each entry in the rank table is uppersemicontinuous, there is a dense open subset $U \subset X'$ on which the rank table (and hence the permutation array) is constant. Let $\partial X' = X' - U$. By the hypothesis $\sum_i \ell(w^i) = \binom{n}{2}$, the morphism $X' \to \mathcal{F}l_n^d$ (remembering the flags $E_\bullet^1, \ldots, E_\bullet^d$) is generically finite, and as $\dim \partial X' < \dim X' = \dim \mathcal{F}l_n^d$, the preimage of a general point of $\mathcal{F}l_n^d$ (i.e. X) misses $\dim \partial X'$. □

We next describe how to compute the rank table described in Theorem 5.1.

5.1. Permutation array algorithm. We describe the rank table of the general element of the product of bundles (5.1). We will compute

$$\dim\left(E_{x_1}^1 \cap \cdots \cap E_{x_p}^p \cap F_{x_{d+1}}\right)$$

inductively on p, where the base case $p = 0$ is trivial. We assume that the answer is known for $p-1$, and describe the case p. Let $V = E_{x_1}^1 \cap \cdots \cap E_{x_{p-1}}^{p-1}$. This meets flag $F_\bullet$ in a known way (by the inductive hypothesis, say the it lies in the Schubert cell $X_\lambda(F_\bullet)$ in the Grassmannian $G(\dim V, n)$), and $E_\bullet^p$ meets $F_\bullet$ in a known way, corresponding to permutation w^p. As we are considering a general element of the product of bundles, the question boils down to the following: given a general element $[V] \in X_\lambda(F_\bullet) \subset G(\dim V, n)$, and a general element $[E_\bullet^p] \in X_{w^p}(F_\bullet) \subset \mathcal{F}l_n$, how do V, $E_\bullet^p$, and $F_\bullet$ meet, i.e. what is

$$\dim(V \cap E_{x_p}^p \cap F_{x_{d+1}})$$

as x_p and x_{d+1} vary through $\{1, \ldots, n\}$? In other words, we have the data of one $n \times n$ table, containing the entries $\dim(E_{x_p}^p \cap F_{x_{d+1}})$ (the data of w^p; here x_p and x_{d+1} vary through $\{1, \ldots, n\}$), and we wish to fill in the entries of another $n \times n$ table, with entries $\dim(V \cap E_{x_p}^p \cap F_{x_{d+1}})$, where one edge (where $x_p = n$) is known (the data of V).

We now address this linear algebra problem. We fix $E_\bullet^p$ and $F_\bullet$, and let V vary in $X_\lambda^\circ(F_\bullet)$. Choose a basis $e_1, \ldots, e_n$ of our n-dimensional space, so that $F_i = \langle e_1, \ldots, e_i \rangle$, and E_i^p is the span of the appropriate i basis elements in terms of the permutation array for the permutation w^p, i.e. E_j^p is the span of the e_i's where $i \in \{w_1, \ldots, w_i\}$ and $w = w_0(w^p)^{-1}$. We can assume that $F_\bullet$ is actually in the Schubert cell $X_{w^p}^\circ(E_\bullet^p)$, not just the Schubert variety $X_{w^p}(E_\bullet^p)$: by repeating this discussion with any component of the boundary, we see that such a boundary locus is of strictly smaller dimension. (Again, the interested reader will readily generalize this discussion to the case where X has positive dimension; the generalization of Theorem 5.1 gives a permutation array for the behavior at the general point of each component of X.)

The Schubert cell X_λ corresponds to a subset $\lambda = \{\lambda_1, \ldots, \lambda_{\dim V}\} \subset \{1, \ldots, n\}$, and a general element $[V]$ of $X_\lambda(F_\bullet)$ is spanned by the vectors

$$v_1 = ?e_1 + \cdots + ?e_{\lambda_1} \tag{5.2}$$
$$v_2 = ?e_1 + \cdots + \widehat{?e_{\lambda_1}} + \cdots + ?e_{\lambda_2}, \tag{5.3}$$
$$\vdots \tag{5.4}$$
$$v_{\dim V} = ?e_1 + \cdots + \widehat{?e_{\lambda_1}} + \cdots + \widehat{?e_{\lambda_{\dim V-1}}} + \cdots + ?e_{\lambda_{\dim V}} \tag{5.5}$$

where the non-zero coefficients (the question marks) are chosen generally. Let V_i be the set of indices j such that e_j has non-zero coefficient in v_i.

We wish to compute $\dim(V \cap E^p_j \cap F_k)$ for each $j = x_p$ and $k = x_{d+1}$. This is now a rank calculation: $V \cap F_k$ is the span of those basis elements of V (in (5.2)–(5.5)) involving no basis elements above e_k. We seek the dimension of the intersection of this with E^p_j, which is the span of known standard basis elements indexed by I_j. Therefore,

$$\dim(V \cap E^p_j \cap F_k) = \dim(\text{span}\{v_i : \lambda_i \le k\} \cap \text{span}\{e_{w_1}, \ldots, e_{w_j}\}) \tag{5.6}$$

where $w = w_0(w^p)^{-1}$ as above. This dimension is the corank of the matrix whose rows are determined by the given basis of $V \cap F_{x_{d+1}}$ and the basis of $E^p_{x_p}$. This can be computed "by eye" as follows. We then look for k columns, and more than k of the first $\dim V$ rows each of whose question marks all appear in the chosen k columns. Whenever we find such a configuration, we erase all but the first k of those rows — the remaining rows are dependent on the first k. The number of rows of the matrix remaining after this operation is the rank of the matrix, and the number of erased rows is the corank.

Thus we have described how to compute the rank table of the general element of the product of bundles (5.1).

One interesting problem in Schubert calculus is to determine efficiently if a structure constant for $H^*(G/B, \mathbb{Z})$ is zero, or equivalently if X is empty. In the case of the Grassmannian manifold, non-empty Schubert problems are related to triples of eigenvalues satisfying Horn's recurrence and Knutson-Tao honeycombs [Knutson and Tao, 2001]. For the flag manifold, both Knutson [Knutson, 2001] and Purbhoo [Purbhoo, 2006] gave a sufficient criteria for vanishing in terms of decent cycling and "root games" respectively. Below we give a criteria for vanishing that is very easy to compute, in fact more efficient than Knutson or Purbhoo's result, however, less comprehensive. As evidence that our criteria is more efficient, we give a pseudo random example in S_{15} which was computed in a few seconds on by a computer. This technique has been extended recently in [Ardila and Billey, 2006] by considering a matroid on the 1-dimensional spans $E^1_{x_1} \cap \cdots \cap E^d_{x_d}$.

COROLLARY 5.1. *Let P be the permutation array whose rank table coincides with the table constructed by the algorithm in Section 5.1 for a given collection of permutations $w^1, \ldots, w^d$ such that $\sum_i \ell(w^i) = \binom{n}{2}$. Let*

P_n be the projection of P onto the first d coordinates following the notation of Theorem 3.3. If $P_n \neq T_{n,d}$, then X is the empty set.

When $d = 4$, this corollary can often be used to detect when the coefficients $c_{u,v}^w$ are zero in Equation 2.2. This criterion catches 7 of the 8 zero coefficients in 3 dimensions, 373 of the 425 in 4 dimensions, and 28920 of the 33265 in dimension 5. The dimension 3 case missed by this criterion is presumably typical of what the criterion fails to see: there are no 2-planes in 3-space containing three general 1-dimensional subspaces. However, given a 2-plane V, three general flags with 1-subspaces contained in V are indeed transverse.

The corollary and algorithm are efficient to apply. For example, consider the following three permutations (anagrams) of the name "Richard P. Stanley" in S_{15}.[1] To interpret these phases as permutations, only the letters count — not spaces or punctuation — the permutation is not case-sensitive, and repeated letters are listed in their original order in the name which is also the identity element.

- $u =$ A Children's Party
- $v =$ Hip Trendy Rascal
- $w =$ Raid Ranch Let Spy.

Using a computer, we can easily compute P_n corresponding to $X = X_u \cap X_v \cap X_w$:

$$P_{15} = \begin{bmatrix}
 & & & & & & & & & & & & & & 11 \\
 & & & & & & & & & & & & & 14 & \\
 & & & & & & & & & & & & 14 & 11 & 3 \\
 & & & & & & & & & & & 15 & & & \\
 & & & & & & & & & & 15 & 14 & 11 & 5 & \\
 & & & & & & & & & 15 & 14 & 11 & 6 & & \\
 & & & & & & & & 15 & 8 & & & & & \\
 & & & & & & & 15 & 14 & & 13 & & & & \\
 & & & & & & 15 & 14 & 8 & 4 & & & & & \\
 & & & & & 15 & 14 & 13 & & & 12 & & & & \\
 & & & & 15 & 14 & 13 & 8 & 3 & & & & & & 2 \\
 & & & 15 & 14 & 13 & 12 & & & & 11 & 9 & & & \\
 & & 15 & 8 & & & & 2 & & & & & & & 1 \\
 & 15 & 13 & & & 11 & & & & & 10 & & & & \\
15 & 13 & 8 & 7 & & & & & & & & & & &
\end{bmatrix}.$$

Clearly, $P_{15} \neq T_{15,3}$ so X is empty and $c_{u,v}^{w_0 w} = 0$.

The array $T_{n,d}$ is an antichain under the dominance order on $[n]^d$ so each element $x \in T_{n,d}$ corresponds with a principle subarray $T_{n,d}[x] = \{x\}$ consisting of a single element. Therefore, each $x \in T_{n,d}$ corresponds with a 1-dimensional vector space $E^1_{x_1} \cap E^2_{x_2} \cap \cdots \cap E^{d-1}_{x_{d-1}} \cap E^d_i$. Define

$$V(E^1_\bullet, \ldots, E^d_\bullet) = \{v_x \mid x \in T_{n,d}\}$$

to be a collection of non-zero vectors chosen such that $v_x \in E^1_{x_1} \cap E^2_{x_2} \cap \cdots \cap E^d_{x_d}$. These lines will provide a "skeleton" for the given Schubert problem.

[1]The name Richard P. Stanley has an unusually high number of interesting anagrams. Stanley has a long list of such anagrams on his office door. They are also available on his homepage by clicking on his name.

THEOREM 5.2. *Let* $X = X_{w^1}(E^1_\bullet) \cap X_{w^2}(E^2_\bullet) \cap \cdots \cap X_{w^d}(E^d_\bullet)$ *be a 0-dimensional intersection, with* $E^1_\bullet, \ldots, E^d_\bullet$ *general. Let* $P \subset [n]^{d+1}$ *be the unique permutation array associated to this intersection. Then polynomial equations defining* X *can be determined simply by knowing* P *and* $V(E^1_\bullet, \ldots, E^d_\bullet)$.

To prove the theorem, we give an algorithm for constructing the promised equations in terms of the data in P and $V(E^1_\bullet, \ldots, E^d_\bullet)$. Then we explain how to construct all the flags in X from the solutions to the equations.

Proof. Let $P_1, \ldots, P_n$ be the sequence of permutation arrays in $[n]^d$ used to obtain P in the EL-Algorithm in Theorem 3.3. If $F_\bullet \in X$, then by Corollary 3.1 P_i is the unique permutation array encoding $\dim(E^1_{x_1} \cap E^2_{x_2} \cap \cdots \cap E^d_{x_d} \cap F_i)$. Furthermore, for each $x \in P_i, 1 \leq i \leq n$, choose a representative vector in the corresponding intersection, say $v^i_x \in E^1_{x_1} \cap E^2_{x_2} \cap \cdots \cap E^d_{x_d} \cap F_i$. Define

$$V_i = \{v^i_y \mid y \in P_i\}$$
$$V_i[x] = \{v^i_y \mid y \in P_i[x]\}.$$

More specifically, choose the vectors v^i_x so that $v^i_x \notin \mathrm{Span}(V_i[x] \setminus \{v^i_x\})$ since the rank function must increase at position x. Therefore, we have

$$\mathrm{v.rk}(V_i[x]) = \mathrm{rk}(P_i[x]) \tag{5.7}$$

for all $x \in [n]^d$ and all $1 \leq i \leq n$ where $\mathrm{v.rk}(S)$ is the dimension of the vector space spanned by the vectors in S. These rank conditions define X.

Let $V_n = V(E^1_\bullet, \ldots, E^d_\bullet)$ be the finite collection of vectors in the case $i = n$. Given V_{i+1}, P_{i+1} and P_i, we compute

$$V_i = \{v^i_x \mid x \in P_i\}$$

recursively as follows. If $x \in P_i \cap P_{i+1}$ then set

$$v^i_x = v^{i+1}_x.$$

If $x \in P_i \setminus P_{i+1}$ and $y, \ldots, z$ is a *basis set* for $P_{i+1}[x]$, i.e. $v^{i+1}_y, \ldots, v^{i+1}_z$ are independent and span the vector space generated by all v^{i+1}_w with $w \in P_{i+1}[x]$, then set

$$v^i_x = c^i_y v^i_y + \cdots + c^i_z v^i_z \tag{5.8}$$

where $c^i_y, \ldots, c^i_z$ are indeterminate. Now the same rank equations as in (5.7) must hold. In fact, it is sufficient in a 0-dimensional variety X to require only

$$\mathrm{v.rk}\{v^i_y \mid y \in P_i[x]\} \leq \mathrm{rk}(P_i[x]) \tag{5.9}$$

for all $x \in [n]^d$ and all $1 \leq i \leq n$. Let $\mathrm{minors}_k(M)$ be the set of all $k \times k$ determinantal minors of a matrix M. Let $M(V_i[x])$ be the matrix whose rows are given by the vectors in $V_i[x]$. Then, the rank conditions (5.9) can be rephrased as

$$\text{minors}_{rk(P_i[x])+1}(M(V_i[x])) = 0 \tag{5.10}$$

for all $1 \leq i < n$ and $x \in [n]^d$ such that $\sum x_i > (d-1)n$.

For each set of solutions S to the equations in (5.10), we obtain a collection of vector sets by substituting solutions for the indeterminates in the formulas (5.8) for the vectors. Note, these "solutions" may be written in terms of other variables so at an intermediate point in the computation, there could potentially be an infinite number of solutions. We further eliminate variables whenever a vector depends only on one variable c^i_y by setting it equal to any nonzero value which does not force another $c^j_z = 0$. If ever a solution implies $c^j_z = 0$, then the choice of $E^1_\bullet, \dots, E^d_\bullet$ was not general. Let $V^S_1, \dots V^S_n$ be the final collection of vector sets depending on the solutions S. Since X is 0-dimensional, if $V^S_1, \dots V^S_n$ depends on any indeterminate then $E^1_\bullet, \dots, E^d_\bullet$ was not general. Let F^S_i be the span of the vectors in V^S_i. Then the flag $F^S_\bullet = (F^S_1, \dots, F^S_n)$ satisfies all the rank conditions defining $X = X^\circ_P(E^1_\bullet, \dots, E^d_\bullet)$. Hence, $F^S_\bullet \in X$. □

COROLLARY 5.2. *The equations appearing in* (5.10) *provide a test for determining if $E^1_\bullet, \dots, E^d_\bullet$ is sufficiently general for the given Schubert problem. Namely, the number of flags satisfying the equations* (5.10) *is the generic intersection number if each indeterminate c^i_z takes a nonzero value, and the solution space determined by the equations is 0-dimensional.*

REMARK 5.1. Theorem 5.2 has two clear advantages over a naive approach to intersecting Schubert varieties. First, we have reduced the computational complexity for finding all solutions to certain Schubert problems. See Section 5.2 for a detailed analysis. Second, we see the permutation arrays as a complete flag analog of the checkerboards in the geometric Littlewood-Richardson rule of [Vakil, 2006a]. More specifically, checker boards are two nested $[n]^2$ permutation arrays. A permutation array P can be thought of as n nested permutation arrays $P_1, P_2, \dots, P_n$ using the notation in Theorem 3.3. Then the analog of the initial board in the checker's game would be the unique $[n]^2$ permutation array corresponding to two permutations u and v, the final boards in the tree would encode the permutations w such that $c^w_{uv} \neq 0$ in (2.2). The "legal moves" from level i to level $i+1$ can be determined by degenerations in specific cases solving the equations in Theorem 5.2, but we don't know a general rule at this time. A two-step version of such a rule is given in [Coskun], see also [Coskun and Vakil].

5.2. Algorithmic complexity. It is well known that solving Schubert problems are "hard". To our knowledge, no complete analysis of the algorithmic complexity is known. We will attempt to show that the approach outlined in Theorem 5.2 typically reduces the number of variables introduced into the system, while unfortunately increasing the number of rank conditions. Therefore, the entire process is still exponential as both n and d grow large.

For a fixed n and d, the following naive approach would imply that a typical Schubert problem would require one to consider $d \cdot n^2$ rank conditions in n^2 variables. First, consider an arbitrary flag $F_\bullet \in \mathcal{F}l_n$. In terms of a fixed basis, $\{e_1, \ldots, e_n\}$, one could give an ordered basis for $F_\bullet$ with n^2 variable coefficients. Then for each permutation w^i for $1 \leq i \leq d$, the condition that $F_\bullet \in X_{w^i}(E^i)$ is equivalent to n^2 rank conditions by definition (2.1). Each rank condition, can be checked via determinantal equations on matrices with entries among the n^2 variables.

One could easily improve the naive computations in two ways:

1. Assume $F_\bullet \in X_{w^1}(E^1)$. Then one would need at most $\binom{n}{2}$ variables and only $(d-1)n^2$ additional rank conditions.
2. Second, some of the rank conditions in (2.1) are redundant. One only needs to check the conditions for pairs in Fulton's essential set [Fulton, 1991]. Eriksson and Linusson [Eriksson and Linusson, 1995] have shown that the average size of the essential set is $\frac{1}{36}n^2$. However, this does not significantly reduce the number of rank equations on average or in the worst case.

In our approach, the number of rank conditions grows like n^d, i.e. polynomial in n for a fixed d but exponential in d. We have succeeded in solving many Schubert problems for $n = 6$ and $d = 3$ using this approach. There are Schubert problems for $n = 8$ and $d = 3$ for which our code in Maple cannot solve the associated system of equations. Computing the unique permutation array associated to a collection of permutations is relatively quick. In the next section we give an example with $n = 15$ and $d = 3$ which was calculated in just a few seconds. Examples with $n = 25$ and $d = 3$ take just over 1 minute.

The main advantage of our approach is that variables are only introduced as necessary. In order to minimize the number of variables, we recommend solving the equations in a particular order. First, it is useful to solve all equations pertaining to V_{i+1} before computing the initial form of the vectors in V_i. Second, we have found that proceeding through all $x \in [n]^d$ such that $\sum x_i > (d-1)n$ in lexicographic order works well, with the additional caveat that if $P_i[x] = \{x\}$ then the matrix M with rows given by the vectors indexed by $\{x\} \cup (P_i \cap P_{i+1})$ must have rank at most i. Solve all of the determinantal equations implying the rank condition $\text{v.rk}(V_i[x]) = \text{rk}(P_i[x])$ simultaneously and substitute each solution back into the collection of vectors before considering the next rank condition. The second point is helpful because we solve all rank i equations before considering the rank $i+1$ equations.

The following table gives the number of free variables necessary for solving *all* Schubert problems with $n = 3, 4, 5$ and $d = 3$. Row n and column i gives the number of Schubert problems for that n requiring i free variables.

	0	1	2	3	4	5
$n = 3$	8	1	0	0	0	0
$n = 4$	176	23	11	1	0	0
$n = 5$	10639	910	585	457	135	0

For $n = 6$, all examples computed so far (over 10,000) require at most 5 free variables.

It is well known in that solving more equations with fewer variables is not necessarily an improvement. More experiments are required to characterize the "best" method of computing Schubert problems. We are limited in experimenting with this solution technique to what a symbolic programming language like Maple can do in a reasonable period of time. The examples in the next section will illustrate how this technique is useful in keeping both the number of variables and the complexity of the rank equations to a minimum.

6. The key example: triple intersections. We now implement the algorithm of the previous section in an important special case. Our goal is to describe a method for directly identifying all flags in $X = X_u(E^1_\bullet) \cap X_v(E^2_\bullet) \cap X_w(E^3_\bullet)$ when $\ell(u) + \ell(v) + \ell(w) = \binom{n}{2}$ and $E^1_\bullet$, $E^2_\bullet$, and $E^3_\bullet$ are in general position. This gives a method for computing the structure constants in the cohomology ring of the flag variety from equations (2.2) and (2.4) .

There are two parts to this algorithm. First, we use Algorithm 5.1 to find the unique permutation array $P \subset [n]^4$ with position vector (u, v, w) such that $P_n = T_{n,3}$. Second, given P we use the equations in (5.10) to find all flags in X.

As a demonstration, we explicitly compute the flags in X in two cases. For convenience, we work over $\mathbb{C}$, but of course the algorithm is independent of the field. In the first there is just one solution which is relatively easy to see "by eye". In the second case, there are two solutions, and the equations are more complicated. The algorithm has been implemented in Maple and works well on examples where $n \leq 8$.

EXAMPLE 1. Let $u = (1,3,2,4)$, $v = (3,2,1,4)$, $w = (1,3,4,2)$. The sum of their lengths is $1+3+2 = 6 = \binom{n}{2}$. The unique permutation array $P \in [4]^4$ determined by Algorithm 5.1 consists of the following dots:

$$\begin{array}{ccccc} (4421) & (4142) & (2442) & (4233) & (3243) \\ (3433) & (4414) & (4324) & (3424) & (3334) \\ (2434) & (2344) & (1444) & & \end{array}$$

The EL-algorithm produces the following list of permutation arrays P_1, P_2, P_3, P_4 in $[4]^3$ corresponding to P:

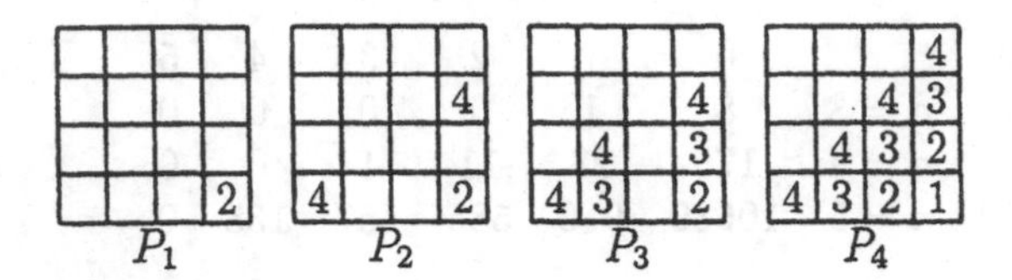

Notice that P_4 is the transverse permutation array $T_{4,3}$. Notice also how to read u, v, and w from $P_1, \ldots, P_4$: P_i has one less row than P_{i+1}; listing these excised rows from right to left yields u. Similarly, listing the excised *columns* from right to left yields v, and listing the excised *numbers* from right to left yields w (see the example immediately above).

We want to specify three transverse fixed flags $E^1_\bullet$, $E^2_\bullet$, $E^3_\bullet$. It will be notationally convenient to represent a vector $v = (v_1, \ldots, v_n)$ by the polynomial $v_1 + v_2 x + \cdots + v_n x^{n-1}$. We choose three flags, or equivalently three "transverse" ordered bases, as follows:

$$\begin{aligned} E^1_\bullet &= \langle 1, x, x^2, x^3 \rangle \\ E^2_\bullet &= \langle x^3, x^2, x, 1 \rangle \\ E^3_\bullet &= \langle (x+1)^3, (x+1)^2, (x+1), 1 \rangle. \end{aligned}$$

We will show that the only flag in $X_u(E^1_\bullet) \cap X_v(E^2_\bullet) \cap X_w(E^3_\bullet)$ is

$$F_\bullet = \langle 2 + 3x - x^3,\ x^3,\ x^2,\ 1 \rangle. \tag{6.1}$$

For each element (i, j, k) in P_4, we choose a vector in the corresponding 1-dimensional intersection $E^1_i \cap E^2_j \cap E^3_k \cap F_4$ and put it in position (i, j) in the matrix below:

$$V(E^1_\bullet, E^2_\bullet, E^3_\bullet) = V_4 = \begin{bmatrix} 0 & 0 & 0 & 1 \\ 0 & 0 & x & x+1 \\ 0 & x^2 & x(x+1) & (x+1)^2 \\ x^3 & x^2(x+1) & x(x+1)^2 & (x+1)^3 \end{bmatrix}.$$

In P_3, every element in the 4th column is covered by a subset in the antichain removed from P_4. This column adds only one degree of freedom so we establish V_3 by adding only one variable in position $(2, 4)$ and solving all other rank two equations in terms of this one:

$$V_3 = \begin{bmatrix} 0 & 0 & 0 & 0 \\ 0 & 0 & 0 & (1+x) + cx \\ 0 & x^2 & 0 & 1 + x + (1+c)x(1+x) \\ x^3 & x^2(x+1) & 0 & (x+1)^2 + cx(x+1)^2 \end{bmatrix}.$$

According to equation (5.8) the entry in position $(4, 2)$ can have two indeterminates: $b(1+x) + cx$, where $b, c \neq 0$. As any two linearly dependent ordered pairs (b, c) yield the same configuration of subspaces, we may normalize b to 1.

Once V_3 is determined, we find the vectors in V_2. In P_2, every element is contained in P_3, so V_2 is a subset of V_3:

$$V_2 = \begin{bmatrix} 0 & 0 & 0 & 0 \\ 0 & 0 & 0 & 1+(2+c)x \\ 0 & 0 & 0 & 0 \\ x^3 & 0 & 0 & (x+1)^2 + cx(x+1)^2 \end{bmatrix}.$$

The rank of P_2 is 2, so all 3×3 minors of the following matrix must be zero:

$$\begin{pmatrix} 0 & 0 & 0 & 1 \\ 1 & 2+c & 0 & 0 \\ 1 & 2+c & 1+2c & c \end{pmatrix}.$$

In particular, $1 + 2c = 0$, so the only solution is $c = -\frac{1}{2}$. Substituting for c, we have

$$V_2^S = \begin{bmatrix} 0 & 0 & 0 & 0 \\ 0 & 0 & 0 & 1+\frac{3}{2}x \\ 0 & 0 & 0 & 0 \\ x^3 & 0 & 0 & (x+1)^2 - \frac{1}{2}x(x+1)^2 \end{bmatrix}.$$

Finally P_1 is contained in P_2, so V_1^S contains just the vector

$$v^1_{(4,4,2)} = v^2_{(4,4,2)} = (x+1)^2 - \frac{1}{2}x(x+1)^2 = \frac{1}{2}(2 + 3x - x^3).$$

Therefore, there is just one solution, namely the flag spanned by the collections of vectors $V_1^S, V_2^S, V_3^S, V_4^S$ which is equivalent to the flag in (6.1).

If we choose an arbitrary general collection of three flags, we can always change bases so that we have the following situation:

$$\begin{aligned} E^1_\bullet &= \langle 1, x, x^2, x^3 \rangle \\ E^2_\bullet &= \langle x^3, x^2, x, 1 \rangle \\ E^3_\bullet &= \langle a_1 + a_2 x + a_3 x^2 + x^3,\, b_1 + b_2 x + x^2,\, c_1 + x,\, 1 \rangle. \end{aligned}$$

Using these coordinates, the same procedure as above will produce the unique solution

$$F_\bullet = \langle (a_1 - a_3 b_1) + (a_2 - b_2 a_3)x - x^3,\, x^3,\, x^2,\, 1 \rangle.$$

EXAMPLE 2. This example is of a Schubert problem with multiple solutions. Let $u = (1,3,2,5,4,6)$, $v = (3,5,1,2,4,6)$, $w = (3,1,6,5,4,2)$. If P is the unique permutation array in $[6]^4$ determined by Algorithm 5.1 for u, v, w then the EL-algorithm produces the following list of permutation arrays $P_1, \ldots, P_6$ in $[6]^3$ corresponding to P:

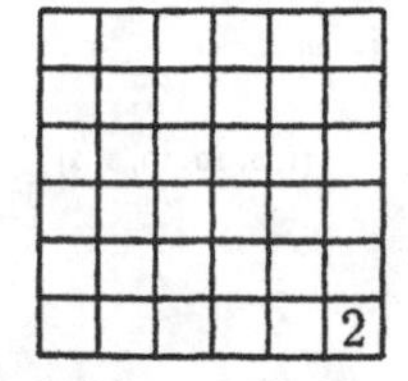

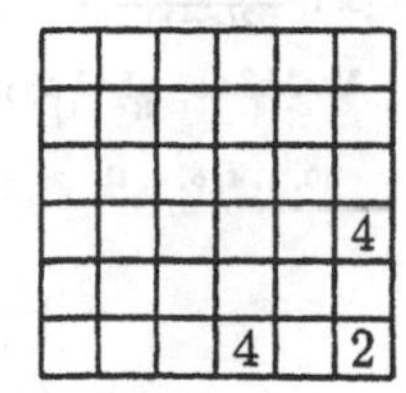

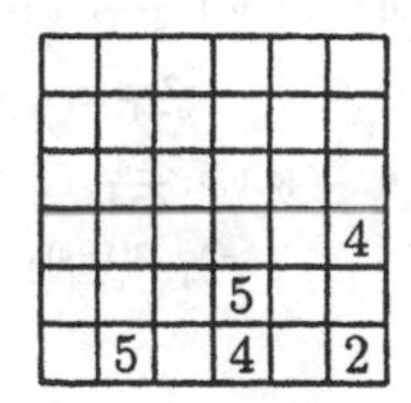

					6
			6		4
	6		5		
6	5		4		2

					6
				6	5
			6	5	4
	6		5	4	2
6	5		4	2	1

					6
				6	5
			6	5	4
		6	5	4	3
	6	5	4	3	2
6	5	4	3	2	1

We take the following triple of fixed flags:

$$\begin{aligned} E^1_\bullet &= \langle 1, x, \ldots, x^5 \rangle \\ E^2_\bullet &= \langle x^5, \ldots, x, 1 \rangle \\ E^3_\bullet &= \langle (1+x)^5, (1+x)^4, \ldots, 1 \rangle \end{aligned}$$

The third flag is clearly not chosen generally but leads to two solutions to this Schubert problem which is the generic number of solutions. We prefer to work with explicit but simple numbers here to demonstrate the computation without making the formulas too complicated.

The vector table associated to P_6 is easily determined by Pascal's formula:

$$\begin{bmatrix} [] & [] & [] & [] & [] & [1,0,0,0,0,0] \\ [] & [] & [] & [] & [0,1,0,0,0,0] & [1,1,0,0,0,0] \\ [] & [] & [] & [0,0,1,0,0,0] & [0,1,1,0,0,0] & [1,2,1,0,0,0] \\ [] & [] & [0,0,0,1,0,0] & [0,0,1,1,0,0] & [0,1,2,1,0,0] & [1,3,3,1,0,0] \\ [] & [0,0,0,0,1,0] & [0,0,0,1,1,0] & [0,0,1,2,1,0] & [0,1,3,3,1,0] & [1,4,6,4,1,0] \\ [0,0,0,0,0,1] & [0,0,0,0,1,1] & [0,0,0,1,2,1] & [0,0,1,3,3,1] & [0,1,4,6,4,1] & [1,5,10,10,5,1] \end{bmatrix}$$

The vector table associated to P_5 has one degree of freedom. The vector in position $(3,5)$ is freely chosen to be $x + cx^2$. Then for all other points in $P_5 \setminus P_6$ we can solve a rank 2 equation which determines the corresponding vector in terms of c. Therefore, V_5 becomes

$$\begin{bmatrix} [] & [] & [] & [] & [] & [] \\ [] & [] & [] & [] & [] & [1, \frac{10(c-1)}{3c}, 0, 0, 0, 0] \\ [] & [] & [] & [] & [0, 1, c, 0, 0, 0] & [1, \frac{-13+10c}{3(c-1)}, \frac{-10+7c}{3(c-1)}, 0, 0, 0] \\ [] & [] & [] & [0, 0, 1, \frac{-6}{c-4}, 0, 0] & [0, 1, \frac{-4+7c}{2+c}, \frac{6(c-1)}{2+c}, 0, 0] & [1, 6\frac{-5+3c}{-8+5c}, \frac{3(-12+7c)}{-8+5c}, \frac{2(-7+4c)}{-8+5c}, 0, 0] \\ [] & x^4 & [] & [0, 0, 1, \frac{-6}{c-4}, \frac{-2-c}{c-4}, 0] & [0, 1, \frac{3(-4+3c)}{2(c-1)}, \frac{3(-3+2c)}{c-1}, \frac{-8+5c}{2(c-1)}, 0] & [1, 4, 6, 4, 1, 0] \\ x^5 & x^4 + x^5 & [] & [0, 0, 1, \frac{-6}{c-4}, \frac{-3c}{c-4}, \frac{2(1-c)}{c-4}] & [0, 1, 4, 6, 4, 1] & [1, 5, 10, 10, 5, 1] \end{bmatrix}$$

Every vector in V_4 appears in V_5, but now some of them are subject to new rank conditions:

$$\begin{bmatrix} [] & [] & [] & [] & [] & [] \\ [] & [] & [] & [] & [] & [1, \frac{10(c-1)}{3c}, 0, 0, 0, 0] \\ [] & [] & [] & [] & [] & [] \\ [] & [] & [] & [0, 0, 1, \frac{-6}{c-4}, 0, 0] & [] & [1, \frac{6(-5+3c)}{-8+5c}, \frac{3(-12+7c)}{-8+5c}, \frac{2(-7+4c)}{-8+5c}, 0, 0] \\ [] & x^4 & [] & [0, 0, 1, \frac{-6}{c-4}, -\frac{2+c}{c-4}, 0] & [] & [] \\ x^5 & x^4+x^5 & [] & [0, 0, 1, \frac{-6}{c-4}, \frac{-3c}{c-4}, \frac{-2c+2}{c-4}] & [] & [1, 4+d, 6+4d, 4+6d, 1+4d, d] \end{bmatrix}$$

In particular, the top 3 vectors should span a two-dimensional subspace. This happens if the following matrix has rank 2:

$$\begin{bmatrix} 1 & \frac{10(c-1)}{3c} & 0 & 0 & 0 & 0 \\ 1 & \frac{6(-5+3c)}{-8+5c} & \frac{3(-12+7c)}{-8+5c} & \frac{2(-7+4c)}{-8+5c} & 0 & 0 \\ 0 & 0 & 1 & \frac{-6}{c-4} & 0 & 0 \end{bmatrix}$$

or equivalently if the following nontrivial minors of the matrix are zero

$$\left[\frac{4(10c+c^2-20)}{3c(-8+5c)}, \frac{-8(10c+c^2-20)}{(-8+5c)(c-4)c}, \frac{-8(10c+c^2-20)}{(-8+5c)(c-4)}, \frac{-8(c-1)(10c+c^2-20)}{3(-8+5c)(c-4)c}\right].$$

All rank 3 minors will be zero if $c^2+10c-20=0$, or $c=-5\pm 3\sqrt{5}$. Plugging each solution for c into the vectors gives the two solutions $V_4^{S_1}$ and $V_4^{S_2}$. For example, using $c=-5+3\sqrt{5}$ and solving a single rank 2 equation involving d gives:

$$\begin{bmatrix} [] & [] & [] & [] & [] & [] \\ [] & [] & [] & [] & [] & [1, 10\frac{2+\sqrt{5}}{5+3\sqrt{5}}, 0, 0, 0, 0] \\ [] & [] & [] & [] & [] & [] \\ [] & [] & [] & [0, 0, 1, \frac{2}{(3+\sqrt{5})}, 0, 0] & [] & [1, 2\frac{20+9\sqrt{5}}{11+5\sqrt{5}}, \frac{47+21\sqrt{5}}{11+5\sqrt{5}}, 2\frac{9+4\sqrt{5}}{11+5\sqrt{5}}, 0, 0] \\ [] & x^4 & [] & [0, 0, 1, \frac{2}{(3+\sqrt{5})}, -\frac{1+\sqrt{5}}{3+\sqrt{5}}, 0] & [] & [] \\ x^5 & x^4+x^5 & [] & [0, 0, 1, \frac{2}{(3+\sqrt{5})}, -\frac{5+3\sqrt{5}}{3+\sqrt{5}}, \frac{-2(2+\sqrt{5})}{3+\sqrt{5}}] & [] & [1, \frac{5+\sqrt{5}}{2}, 2\sqrt{5}, -5+3\sqrt{5}, -5+2\sqrt{5}, \frac{-3+\sqrt{5}}{2}] \end{bmatrix}$$

The remaining vectors in $V_1^{S_1}, V_2^{S_1}, V_3^{S_1}$ will be a subset of $V_4^{S_1}$ so no further equations need to be solved, and similarly for $V_4^{S_2}$.

7. Monodromy and Galois groups. The monodromy group of a problem in enumerative geometry captures information reflecting three aspects of algebraic geometry: geometry, algebra, and arithmetic. Informally, it is the symmetry group of the set of solutions. Three more precise interpretations are given below. Historically, these groups were studied since the nineteenth century [Jordan, 1870, Dickson et al., 1916, Weber, 1941]; modern interest probably dates from a letter from Serre to Kleiman in the seventies (see the historical discussion in the survey article [Kleiman, 1987, p. 325]). Their modern foundations were laid by Harris [Harris, 1979]; among other things, he showed that the monodromy group of a problem is equivalent to the Galois group of the equations defining it.

These groups are difficult to compute in general, and indeed they are known for relatively few enumerative problems. In this section, we use the computation of explicit algebraic solutions to Schubert problems (along with a criterion from [Vakil, 2006b]) to give a method to compute many such groups explicitly (when they are "full", or as large as possible), and to give an experimental method to compute groups in other cases.

It is most interesting to exhibit cases where the Galois/monodromy group is unexpectedly small. Indeed, Harris writes of his calculations:

> the results represent an affirmation of one understanding of the geometry underlying each of these problems, in the following sense: in every case dealt with here, the actual structure on the set of solutions of the enumerative problem as determined by the Galois group of the problems, is readily described in terms of standard algebrao-geometric constructions. In particular, in every case in which current theory had failed to discern any intrinsic structure on the set of solutions — it is proved here — there is in fact none. [Harris, 1979, p. 687-8]

We exhibit an example of a Schubert problem whose Galois/monodromy group experimentally appears to be smaller than expected — it is the dihedral group $D_4 \subset S_4$. This is the first example in which current theory fails to discern intrinsic structure. Examples of "small" Galois groups were given in [Vakil, 2006b, Sect. 5]; but there an explanation had already been given by Derksen. Here, however, we have a mystery: We do not understand geometrically why the group is D_4. (However, see the end of this section for a conjectural answer.)

We now describe the three interpretations of the Galois/monodromy group for a Schubert problem. The definition for a general problem in enumerative geometry is the obvious generalization; see [Harris, 1979] for a precise definition, and for the equivalence of *(A)* and *(B)*. See [Vakil, 2006b, Sect. 2.9] for more discussion.

(A) Geometry. Begin with m general flags; suppose there are N solutions to the Schubert problem (i.e. there are N flags meeting our m given flags in the specified manner). Move the m flags around in such a way

that no two of the solutions ever come together, returning the m flags to their starting positions, and follow the N solutions. The N solutions are returned to their initial positions as a *set*, but the individual N solutions may be permuted. What are the possible permutations? (See the applet `http://lamar.colostate.edu/~jachter/mono.html` for an illustration of this concept.)

(B) Algebra. The m flags are parameterized by $\mathcal{F}l_n^m$. Define the "solution space" to be the subvariety of $\mathcal{F}l_n \times \mathcal{F}l_n^m$ mapping to $\mathcal{F}l_n^m$, corresponding to those flags satisfying the given Schubert conditions. There is one irreducible component X of the solution space mapping dominantly to $\mathcal{F}l_n^m$; the morphism has generic degree N. The Galois/monodromy group is the Galois group of the Galois closure of the corresponding extension of function fields. The irreducibility of X implies that the Galois group G is a transitive subgroup of S_N.

(C) Arithmetic. If the m flags are defined over $\mathbb{Q}$, then the smallest field of definition of a solution must have Galois group that is a subgroup of the Galois/monodromy group G. Moreover, for a randomly chosen set of m flags, the field of definition will have Galois group precisely G with positive probability (depending on the particular problem). The equivalence of this version with the previous two follows from *(B)* by the Hilbert irreducibility theorem, as $\mathcal{F}l_n^m$ is rational ([Lang, 1983, Sect. 9.2], see also [Serre, 1989, Sect. 1.5] and [Cohen, 1981]). We are grateful to M. Nakamaye for discussions on this topic.

Given any enumerative problem with N solutions, we see that the Galois/ monodromy group is a subgroup of S_N; it is well-defined up to conjugacy in S_N. As the solution set should be expected to be as symmetric as possible, one should expect it to be as large as possible; it should be S_N unless the set of solutions has some additional geometric structure.

For example, in [Harris, 1979], Harris computed several Galois/ monodromy groups, and in each case they were the full symmetric group, unless there was a previously known geometric reason why the group was smaller. The incidence relations of the 27 lines on a smooth cubic surface prevent the corresponding group from being two-transitive. There exist two of the 27 lines that intersect, and there exist another two that do not. These incidence relations can be used to show that the Galois/monodromy group must be contained in the reflection group $W(E_6) \subset S_{27}$, e.g. [Manin, 1974, Sects. 25, 26] or [Hartshorne, 1977, Prob. V.4.11]; Harris shows that equality holds [Harris, 1979, III.3].

Other examples can be computed based on permutation arrays.

COROLLARY 7.1. *The explicit equations defining a Schubert problem in Theorem 5.2 can be used to determine the Galois/monodromy group for the problem as well.*

As a toy example, we see that the monodromy group for Example 2 is S_2, as there are two solutions to the Schubert problem, and the only transitive subgroup of S_2 is S_2 itself. Algebraically, this corresponds to the fact that the roots of the irreducible quadratic $c^2 + 10c - 20$ in example 2 generate a Galois extension of $\mathbb{Q}$ with Galois group S_2.

Unfortunately, the calculations of monodromy groups for flag varieties becomes computationally infeasible as $n \to 10$ where the number of solutions becomes larger. Therefore, we have considered related problems of computing Schubert problems for the Grassmannian manifolds $G(k, n)$. Here, $G(k, n)$ is the set of k-dimensional planes in $\mathbb{C}^n$. Schubert varieties are defined analogously by rank conditions with respect to a fixed flag. These varieties are indexed by partitions $\lambda = (\lambda_1, \ldots, \lambda_k)$ where $\lambda_1 \geq \cdots \geq \lambda_k \geq 0$. The permutation arrays work equally well for keeping track of the rank conditions for intersecting Schubert varieties in the Grassmannian if we replace the condition that a permutation array must have rank n by requiring rank k.

In the case of the Grassmannian, combinatorial criteria were given for the Galois/monodromy group of a Schubert problem to be A_N or S_N in [Vakil, 2006b]. Intersections on the Grassmannian manifold may be interpreted as a special case of intersections on the flag manifold, so our computational techniques apply. We sketch the criteria here, and refer the reader to [Vakil, 2006b] for explicit descriptions and demonstrations.

CRITERION 7.1. **SCHUBERT INDUCTION.** *Given a Schubert problem in the Grassmannian manifold, a choice of geometric degenerations yields a directed rooted tree. The edges are directed away from the root. Each vertex has out-degree between* 0 *and* 2. *The portion of the tree connected to an outward-edge of a vertex is called a* branch *of that vertex. Let N be the number of leaves in the tree.*

- *(i) Suppose each vertex with out-degree two satisfies either (a) there are a different number of leaves on the two branches, or (b) there is one leaf on each branch. Then the Galois/monodromy group of the Schubert problem is A_N or S_N.*
- *(ii) Suppose each vertex with out-degree two has a branch with one leaf. Then the Galois/monodromy group of the Schubert problem is S_N.*
- *(iii) Suppose that each vertex with out-degree two satisfies (a) or (b) above, or (c) there are $m \neq 6$ leaves on each branch, and it is known that the corresponding Galois/monodromy group is two-transitive. Then the Galois/monodromy group is A_N or S_N.*

Part (i) is [Vakil, 2006b, Thm. 5.2], (ii) follows from the proof of [Vakil, 2006b, Thm. 5.2], and (iii) is [Vakil, 2006b, Thm. 5.10]. Criterion (i) seems to apply "almost always". Criterion (ii) applies rarely. Criterion (iii) requires additional information and is useful only in ad hoc circumstances.

The method discussed in this paper of explicitly (algebraically) solving Schubert problems gives two new means of computing Galois groups. The first, in combination with the Schubert induction rule, is a straightforward means of proving that a Galois group is the full symmetric group. The second gives strong experimental evidence (but no proof!) that a Galois group is smaller than expected.

CRITERION 7.2. **CRITERION FOR GALOIS/MONODROMY GROUP TO BE FULL.** *If m flags defined over $\mathbb{Q}$ are exhibited such that the solutions are described in terms of the roots of an irreducible degree N polynomial $p(x)$, and this polynomial has a discriminant that is not a square, then by the arithmetic interpretation (C) above, the Galois/monodromy group is not contained in A_N.*

Hence in combination with the Schubert induction criterion (i), this gives a criterion for a Galois/monodromy group to be the full symmetric group S_N.

(In principle one could omit the Schubert induction criterion: if one could exhibit a single Schubert problem defined over $\mathbb{Q}$ whose Galois group was S_N, then the Galois/monodromy group would have to be S_N as well. However, showing that a given degree N polynomial has Galois group S_N is difficult; our discriminant criterion is immediate to apply.)

The smallest Schubert problem where Criterion 7.1(i) applies but Criterion 7.1(ii) does not is the intersection of six copies of the Schubert variety indexed by the partition (1) in $G(2,5)$ (and the dual problem in $G(3,5)$). Geometrically, it asks how many lines in $\mathbb{P}^4$ meet six planes. When the planes are chosen generally, there are five solutions (i.e. five lines). By satisfying the first criterion we know the Galois/monodromy group is "at least alternating" i.e. either A_N or S_N, but we don't know that the group is S_N. We randomly chose six planes defined over $\mathbb{Q}$. Maple found the five solutions, which were in terms of the solutions of the quintic $101z^5 - 554z^4 + 887z^3 - 536z^2 + 194z - 32$. This quintic has non-square discriminant, so we conclude that the Galois/monodromy group is S_5. As other examples, the Schubert problem $(2)^2(1)^4$ in $G(2,6)$ has full Galois/monodromy group S_6, the Schubert problem $(2)(1)^6$ in $G(2,6)$ has full Galois/monodromy group S_9, and the Schubert problem $(2,2)(1)^5$ in $G(3,6)$ has full Galois/monodromy group S_6. We applied this to many Schubert problems and found no examples satisfying Criterion 7.1(i) or (iii) that did not have full Galois group S_N.

As an example of the limits of this method, solving the Schubert problem $(1)^8$ in $G(2,6)$ is not computationally feasible (it has 14 solutions), so this is the smallest Schubert problem whose Galois/monodromy group is unknown (although Criterion 7.1(i) applies, so the group is A_{14} or S_{14}).

CRITERION 7.3. **PROBABILISTIC EVIDENCE FOR SMALLER GALOIS OR MONODROMY GROUPS.** *If for a fixed Schubert problem, a large number of "random" choices of flags in $\mathbb{Q}^n$ always yield Galois groups contained in a proper subgroup $G \subset S_N$, and the group G is achieved for some choice of Schubert conditions, this gives strong evidence that the Galois/monodromy group is G.*

This is of course not a proof — we could be very unlucky in our "random" choices of conditions — but it leaves little doubt.

As an example, consider the Schubert problem $(2,1,1)(3,1)(2,2)^2$ in $G(4,8)$. There are four solutions to this Schubert problem. When random (rational) choices of the four conditions are taken, Maple always (experimentally!) yields a solution in terms of $\sqrt{a+b\sqrt{c}}$ where a, b, and c are rational. The Galois group of any such algebraic number is contained in D_4: it is contained in S_4 as $\sqrt{a+b\sqrt{c}}$ has at most 4 Galois conjugates, and the Galois closure may be obtained by a tower of quadratic extensions over $\mathbb{Q}$. Thus the Galois group is a 2-subgroup of S_4 and hence contained in a 2-Sylow subgroup D_4.

We found a specific choice of Schubert conditions for which the Galois group of the Galois closure K of $\mathbb{Q}(\sqrt{a+b\sqrt{c}})$ over $\mathbb{Q}$ was D_4. (The numbers a, b, and c are large and hence not included here; the Galois group computation is routine.) Thus we have rigorously shown that the Galois group is at least D_4, hence D_4 or S_4. We have strong experimental evidence that the group is D_4.

Challenge: Prove that the Galois group of this Schubert problem is D_4.

We conjecture that the geometry behind this example is as follows. Given four general conditions, the four solutions may be labeled $V_1, \ldots, V_4$ so that either (i) $\dim(V_i \cap V_j) = 0$ if $i \equiv j \pmod 2$ and $\dim(V_i \cap V_j) = 2$ otherwise, or (ii) $\dim(V_i \cap V_j) = 2$ if $i \equiv j \pmod 2$ and $\dim(V_i \cap V_j) = 0$ otherwise. If (i) or (ii) holds then necessarily $G \neq S_4$, implying $G \cong D_4$.

This example (along with the examples of [Vakil, 2006b, Sect. 5.12]) naturally leads to the following question. Suppose $V_1, \ldots, V_N$ are the solutions to a Schubert problem (with generally chosen conditions). Construct a rank table

$$\left\{ \dim \left(\bigcap_{i \in I} V_i \right) \right\}_{I \subset \{1,\ldots,n\}}.$$

In each known example, the Galois/monodromy group is precisely the group of permutations of $\{1, \ldots, n\}$ preserving the rank table.

Question: Is this always true?

REMARK 7.1. Schubert problems for the Grassmannian varieties were among the first examples where the Galois/monodromy groups may be smaller than expected. The first example is due to H. Derksen; the "hidden geometry" behind the smaller Galois group is clearer from the point of view of quiver theory. Derksen's example, and other infinite families of examples, are given in [Vakil, 2006b, Sect. 5.13–5.15].

8. Acknowledgments. We want to thank Federico Ardila for pointing out a subtle error in an earlier draft of the paper. We would also like to thank Eric Babson, Mike Nakamaye, and Frank Sottile for helpful discussions. Along the way, we have done a lot of experimentation on computers using both free and for-profit software including Allegro Common Lisp and Maple.

REFERENCES

[Ardila and Billey, 2006] F. ARDILA AND S. BILLEY, *Flag arrangements and triangulations of products of simplices*, to appear in Advances in Math.

[Cohen, 1981] S.D. COHEN, *The distribution of Galois groups and Hilbert's irreducibility theorem*, Proc. London Math. Soc. (3) **43** (1981), no. 2, 227–250.

[Coskun] I. COSKUN, *A Littlewood-Richarson rule for the two-step flag varieties*, preprint, 2004.

[Coskun and Vakil] I. COSKUN AND R. VAKIL, *Geometric positivity in the cohomology of homogeneous spaces and generalized Schubert calculus*, arXiv:math.AG/0610538.

[Coxeter and Greitzer, 1967] H.S.M. COXETER AND S.L. GREITZER, *Geometry Revisited*, Math. Ass. of Amer., New Haven, 1967.

[Dickson et al., 1916] L. DICKSON, H.F. BLICHFELDT, AND G.A. MILLER, *Theory and applications of finite groups*, John Wiley, New York, 1916.

[Eisenbud and Saltman, 1987] D. EISENBUD AND D. SALTMAN, *Rank varieties of matrices*, Commutative algebra (Berkeley, CA, 1987), 173–212, Math. Sci. Res. Inst. Publ. 15, Springer, New York, 1989.

[Eriksson and Linusson, 1995] K. ERIKSSON AND S. LINUSSON, *The size of Fulton's essential set*, Sém. Lothar. Combin., 34 (1995), pp. Art. B34l, approx. 19 pages (electronic).

[Eriksson and Linusson, 2000a] K. ERIKSSON AND S. LINUSSON, *A combinatorial theory of higher-dimensional permutation array*, Adv. in Appl. Math. **25** (2000), no. 2, 194–211.

[Eriksson and Linusson, 2000b] K. ERIKSSON AND S. LINUSSON, *A decomposition of* $\mathrm{Fl}(n)^d$ *indexed by permutation arrays*, Adv. in Appl. Math. **25** (2000), no. 2, 212–227.

[Fulton, 1991] W. FULTON, *Flags, Schubert polynomials, degeneracy loci, and determinantal formulas*, Duke Math. J., 65 (1991), pp. 381–420.

[Fulton, 1997] W. FULTON, *Young tableaux, with Applications to Representation Theory and Geometry*, London Math. Soc. Student Texts **35**, Cambridge U.P., Cambridge, 1997.

[Gonciulea and V. Lakshmibai, 2001] N. GONCIULEA AND V. LAKSHMIBAI, *Flag varieties*, Hermann-Actualities Mathématiques, 2001.

[Harris, 1979] J. HARRIS, *Galois groups of enumerative problems*, Duke Math. J. **46** (1979), no. 4, 685–724.

[Hartshorne, 1977] R. HARTSHORNE, *Algebraic Geometry*, GTM 52, Springer-Verlag, New York-Heidelberg, 1977.

[Jordan, 1870] C. JORDAN, *Traité des Substitutions*, Gauthier-Villars, Paris, 1870.

[Kleiman, 1987] S. KLEIMAN, *Intersection theory and enumerative geometry: a decade in review*, in *Algebraic geometry, Bowdoin, 1985*, Proc. Sympos. Pure Math., **46**, Part 2, 321-370, Amer. Math. Soc., Providence, RI, 1987.

[Knutson, 2001] A. KNUTSON,*Descent-cycling in Schubert calculus*, Experiment. Math., **10** (2001), no. 3, 345–353.

[Knutson and Tao, 2001] A. KNUTSON AND T. TAO, *Honeycombs and sums of Hermitian matrices*, Notices Amer. Math. Soc., **48** (2001), 175–186.

[Kumar, 2002] S. KUMAR, *Kac-Moody Groups, Their Flag Varieties and Representation Theory*, Progress in Math., **204**, Birkhäuser, Boston, 2002.

[Lang, 1983] S. LANG, *Fundamentals of Diophantine Geometry*, Springer-Verlag, New York, 1983.

[Lascoux and Schützenberger, 1982] LASCOUX, A. AND M.-P. SCHÜTZENBERGER, *Polynômes de Schubert*, C. R. Acad. Sci. Paris Sér. I Math. **294** (1982), no. 13, 447–450.

[Macdonald, 1991] I.G. MACDONALD, *Notes on Schubert Polynomials*, Publ. du LACIM Vol. 6, Université du Québec à Montréal, Montreal, 1991.

[Magyar, 2005] P. MAGYAR, *Bruhat order for two flags and a Line*, Journal of Algebraic Combinatorics, **21** (2005).

[Magyar and van der Kallen, 1999] P. MAGYAR AND W. VAN DER KALLEN, *The Space of triangles, vanishing theorems, and combinatorics*, Journal of Algebra, **222** (1999), 17–50.

[Manin, 1974] YU. MANIN, *Cubic forms: Algebra, Geometry, Arithmetic*, North-Holland, Amsterdam, 1974.

[Manivel, 1998] L. MANIVEL, *Symmetric Functions, Schubert Polynomials and Degeneracy Loci*, J. Swallow trans. SMF/AMS Texts and Monographs, Vol. **6**, AMS, Providence RI, 2001.

[Mnëv, 1985] N. MNËV, *Varieties of combinatorial types of projective configurations and convex polyhedra*, Dolk. Akad. Nauk SSSR, **283** (6) (1985), 1312–1314.

[Mnëv, 1988] N. MNËV, *The universality theorems on the classification problem of configuration varieties and convex polytopes varieties*, in *Topology and geometry — Rohlin seminar*, Lect. Notes in Math. 1346, Springer-Verlag, Berlin, 1988, 527–543.

[Purbhoo, 2006] K. PURBHOO, *Vanishing and nonvanishing criteria in Schubert calculus*, International Math. Res. Not., Art. ID 24590 (2006), pp.1–38.

[Serre, 1989] J.-P. SERRE, *Lectures on the Mordell-Weil theorem*, M. Waldschmidt trans. F. Viehweg, Braunschweig, 1989.

[Shapiro et al., 1997] B. SHAPIRO, M. SHAPIRO, AND A. VAINSHTEIN, *On combinatorics and topology of pairwise intersections of Schubert cells in SL_n/B*, in *The Arnol'd-Gelfand Mathematical Seminars*, 397–437, Birkhäuser, Boston, 1997.

[Vakil, 2006a] R. VAKIL, *A geometric Littlewood-Richardson rule*, with an appendix joint with A. Knutson, Ann. of Math. (2) **164** (2006), no. 2, 371–421.

[Vakil, 2006b] R. VAKIL, *Schubert induction*, Ann. of Math. (2) **164** (2006), no. 2, 489–512.

[Vakil, 2006c] R. VAKIL, *Murphy's Law in algebraic geometry: Badly-behaved deformation spaces*, Invent. Math. **164** (2006), no. 3, 569–590.

[Weber, 1941] H. WEBER, *Lehrbuch der Algebra*, Chelsea Publ. Co., New York, 1941.

EFFICIENT INVERSION OF RATIONAL MAPS OVER FINITE FIELDS

ANTONIO CAFURE*, GUILLERMO MATERA†, AND ARIEL WAISSBEIN‡

Abstract. We study the problem of finding the inverse image of a point in the image of a rational map $F : \mathbb{F}_q^n \to \mathbb{F}_q^n$ over a finite field $\mathbb{F}_q$. Our interest mainly stems from the case where F encodes a permutation given by some public-key cryptographic scheme. Given an element $y^{(0)} \in F(\mathbb{F}_q^n)$, we are able to compute the set of values $x^{(0)} \in \mathbb{F}_q^n$ for which $F(x^{(0)}) = y^{(0)}$ holds with $O(\mathsf{T}n^{4.38}D^{2.38}\delta \log^2 q)$ bit operations, up to logarithmic terms. Here T is the cost of the evaluation of $F_1, \ldots, F_n$, D is the degree of F and δ is the degree of the graph of F.

Key words. Finite fields, polynomial system solving, public-key cryptography, matrices of fixed displacement rank.

AMS(MOS) subject classifications. 14G05, 68W30, 11T71, 68Q25, 47B35.

1. Introduction. Let $\mathbb{F}_q$ be the finite field of q elements, let $\overline{\mathbb{F}}_q$ denote its algebraic closure and let $\mathbb{A}^n$ denote the n-dimensional affine space $\overline{\mathbb{F}}_q^n$ considered as a topological space endowed with the Zariski topology. Let $X := (X_1, \ldots, X_n)$ be a vector of indeterminates and let $F_1, \ldots, F_n$ be elements of the field $\mathbb{F}_q(X) := \mathbb{F}_q(X_1, \ldots, X_n)$ of rational functions in $X_1, \ldots, X_n$ with coefficients in $\mathbb{F}_q$. We consider the rational map $F : \mathbb{A}^n \to \mathbb{A}^n$ defined as $F(x) := (F_1(x), \ldots, F_n(x))$. Assume that the restriction of F to $\mathbb{F}_q^n$ is a partially-defined mapping $F : \mathbb{F}_q^n \to \mathbb{F}_q^n$, i.e., F is well-defined on a nonempty subset of $\mathbb{F}_q^n$. In such a case, we have that $F : \mathbb{F}_q^n \to \mathbb{F}_q^n$ agrees with an $\mathbb{F}_q$-definable polynomial map F^* on its domain (see, e.g., [25]). Unfortunately, the degrees of the polynomials defining F^* may grow exponentially, which prevents us to replace the rational mapping F with the corresponding polynomial map F^*. In this paper we exhibit an algorithm which, given $y^{(0)} \in F(\mathbb{F}_q^n)$, computes $x^{(0)} \in \mathbb{F}_q^n$ such that $F(x^{(0)}) = y^{(0)}$ holds.

A possible approach to this problem consists in computing the inverse mapping of F, provided that F is polynomially or rationally invertible. This is done in [41], where the authors describe an algorithm for inverting a bijective polynomial map $F : \mathbb{A}^n \to \mathbb{A}^n$ defined over $\mathbb{F}_q$, assuming that F

*Depto. de Matemática, Facultad de Ciencias Exactas y Naturales, Universidad de Buenos Aires, Ciudad Universitaria, Pabellón I, (C1428EHA) Buenos Aires, Argentina. Instituto del Desarrollo Humano, Universidad Nacional de General Sarmiento, J.M. Gutiérrez 1150 (1613) Los Polvorines, Argentina (acafure@dm.uba.ar).

†Instituto del Desarrollo Humano, Universidad Nacional de General Sarmiento, J.M. Gutiérrez 1150 (1613) Los Polvorines, Argentina. CONICET, Argentina (gmatera@ungs.edu.ar).

‡CoreLabs, Core Security Technologies, Humboldt 1967 (C1414CTU) Ciudad de Buenos Aires, Argentina. Doctorado en Ingeniería, ITBA: Av. Eduardo Madero 399 (C1106ACD) Cdad. de Buenos Aires, Argentina (Ariel.Waissbein@corest.com).

is an automorphism of $\mathbb{F}_q[X]^n$ whose inverse has degree $(dn)^{O(1)}$, where d is the maximum of the degrees of the polynomials $F_1, \ldots, F_n$. The algorithm performs $(\mathsf{T}nd)^{O(1)}$ arithmetic operations in $\mathbb{F}_q$, where T is the number of arithmetic operations required to evaluate F. Nevertheless, the assumption of the existence of a polynomial or rational inverse of F with degree polynomially bounded seems to be too restrictive to be useful in practice.

From the cryptographic point of view, the critical problem is that of computing the inverse image of a given point $y^{(0)} \in F(\mathbb{F}_q^n)$ under the map F, rather than that of inverting F itself. In this sense, our problem may be reduced to that of solving polynomial systems over a finite field. Unfortunately, it is well known that solving polynomial systems over a finite field is an NP–complete problem, even with quadratic equations with coefficients in $\mathbb{F}_2$ [19]. This has led to the construction of several multivariate public–key cryptoschemes whose security is based on this difficulty. In fact, since [24] researchers have tried to construct public–key schemes based on this apparent difficulty (see, e.g., [36]), but proposals are typically proved to be weak through *ad hoc* attacks (see, e.g., [35], [29], [43]). This might be seen as an indication that the polynomial systems used in public–key cryptography are not intrinsically difficult to solve, and calls for the study of parameters to measure such difficulty.

In this article we exhibit a probabilistic algorithm that computes the inverse image of a point $y^{(0)} \in F(\mathbb{F}_q^n)$ with a cost which depends polynomially on two geometric parameters: the degree D of the map F and the degree δ of the graph of F.

1.1. Outline of our approach. Let $Y = (Y_1, \ldots, Y_n)$ be a vector of new indeterminates. We consider the Zariski closure of the graph $Y = F(X)$ of the morphism F as an $\mathbb{F}_q$–variety $V \subset \mathbb{A}^{2n}$, i.e., as the set of common zeros in $\mathbb{A}^{2n}$ of a finite set of polynomials in $\mathbb{F}_q[X, Y]$. We suppose that the projection morphism $\pi : V \to \mathbb{A}^n$ defined as $\pi(x, y) = y$ is a dominant map of degree D and denote by δ the degree of V. It turns out that V is an absolutely irreducible variety of dimension n, and the generic fiber of π is finite. With a slight abuse of notation we shall identity $F^{-1}(y^{(0)})$ with the fiber $\pi^{-1}(y^{(0)})$.

In order to compute all the q–rational points of $F^{-1}(y^{(0)})$, i.e., the points in the set $F^{-1}(y^{(0)}) \cap \mathbb{F}_q^n$, we shall deform the system $F(X) = y^{(0)}$ into a system $F(X) = y^{(1)} := F(x^{(1)})$ with a point $x^{(1)}$ randomly chosen in a suitable finite field extension $\mathbb{K}$ of $\mathbb{F}_q$. The deformation is given by the curve $C \subset \mathbb{A}^{n+1}$ defined by the equations $F(X) = y^{(1)} + (S-1)(y^{(1)} - y^{(0)})$. For this purpose, we obtain upper bounds on the degree of the generic condition underlying the choice of $x^{(1)}$, which allows us to determine the cardinality of a finite field extension $\mathbb{K}$ of $\mathbb{F}_q$ where such a random choice is possible with good probability of success (see Section 4.2).

The algorithm computing the set $F^{-1}(y^{(0)}) \cap \mathbb{F}_q^n$ may be divided in three main parts. First, we compute a polynomial $m_S(S, T)$ defining a plane curve birationally equivalent to C. This polynomial is obtained as the solution of a system of linear equations whose matrix is block–Toeplitz. Such a solution is computed by applying an efficient algorithm based on the theory of matrices of fixed displacement rank due to [8] (see Section 5.1). Then, in Section 5.2 we extend the computation of the defining polynomial $m_S(S,T)$ of the plane curve birationally equivalent to C to the computation of the birational inverse itself. Finally, in Section 5.3 we substitute 0 for S in the birational inverse computed in the previous step and recover the 0–dimensional fiber $\pi^{-1}(y^{(0)})$, from which we obtain all the points of $F^{-1}(y^{(0)})$ with coordinates in $\mathbb{F}_q$.

The cost of our algorithm is roughly $O(\mathsf{T}n^{4.38}D^{2.38}\delta\log^2 q)$ bit operations, up to logarithmic terms, where T is the cost of the evaluation of the rational map F. Therefore, we extend and improve the results of [12], which require F to be a polynomial map defining a bijection from $\mathbb{F}_q^n$ to $\mathbb{F}_q^n$. This extension allows us to deal with cryptographic schemes such as the so–called "Tractable Rational Map" cryptosystem (see [42], [43, Section 6]). On the other hand, we observe that, if the hypotheses of [41] hold, then our algorithm meets also the complexity bound $(\mathsf{T}nd)^{O(1)}$ of [41].

As mentioned before, another alternative approach in order to compute one or all the q–rational points of $F^{-1}(y^{(0)})$ could be to apply a general algorithm for finding one or all the q–rational solutions of a given polynomial system over a finite field. Algorithms directly aimed at solving polynomial systems over a finite field are usually based on Gröbner basis computations ([16], [3], [4]), elimination techniques ([23], [10]) or relinearization ([29], [14]). Unfortunately, all these algorithms have worst–case exponential running time, and only [10] achieves a polynomial cost in the Bézout number of the system (which is nevertheless exponential in worst case). Furthermore, it is not clear how these algorithms could profit from the knowledge that a given map has associated geometric parameters of "low" value, as happens in certain cryptographic situations.

Finally, from the cryptographic point of view, we observe that withstanding the differential cryptanalysis of E. Biham and A. Shamir ([6], [15]) has become a *de facto* requirement for any block cipher. On the other hand, there is no "strong" test which allows one to analyze the security of cryptosystems based on the problem of solving multivariate equations. Our algorithm may be considered as a first step in this direction.

2. Notions and notations. Throughout this paper we shall denote by $|A|$ the number of elements of a given finite set A.

Let $\mathbb{K}$ be a finite field extension of $\mathbb{F}_q$, let $\mathbb{A}^n := \mathbb{A}^n(\overline{\mathbb{F}}_q)$ be the n–dimensional $\overline{\mathbb{F}}_q^n$ endowed with its Zariski topology, and let $V \subset \mathbb{A}^n$ be a $\mathbb{K}$–variety, that is, the set of common zeros in $\mathbb{A}^n$ of a finite set of polynomials of $\mathbb{K}[X]$. We denote by $\mathbf{I}(V) \subset \mathbb{K}[X]$ the ideal of the variety V, by $\mathbb{K}[V] := \mathbb{K}[X]/\mathbf{I}(V)$ its coordinate ring and by $\mathbb{K}(V)$ its field of total fractions.

For a given irreducible $\mathbb{K}$–variety $V \subset \mathbb{A}^n$, we define its *degree* $\deg V$ as the maximum number of points lying in the intersection of V with an affine linear variety $L \subset \mathbb{A}^n$ of codimension $\dim V$ for which $|V \cap L|$ is finite. More generally, if $V = C_1 \cup \cdots \cup C_N$ is the decomposition of V into irreducible $\mathbb{K}$–components, we define the degree of V as $\deg V := \sum_{i=1}^{N} \deg C_i$ (cf. [22]).

We say that a $\mathbb{K}$–variety $V \subset \mathbb{A}^n$ is *absolutely irreducible* if it is an irreducible $\overline{\mathbb{F}}_q$–variety.

Let $V \subset \mathbb{A}^n$ be an irreducible $\mathbb{K}$–variety and let $\pi : V \to \mathbb{A}^n$ be a dominant mapping. Then we have an algebraic field extension $\mathbb{K}(\mathbb{A}^n) \hookrightarrow \mathbb{K}(V)$. The degree $\deg \pi$ of π is defined as the degree of the field extension $\mathbb{K}(\mathbb{A}^n) \hookrightarrow \mathbb{K}(V)$. We call a point $y \in \mathbb{A}^n$ a *lifting point* of π if the number of inverse images of y equals the degree of the morphism π.

2.1. Data structures. Algorithms in elimination theory are usually described using the standard dense (or sparse) complexity model, i.e., encoding multivariate polynomials by means of the vector of all (or of all nonzero) coefficients. Taking into account that a generic n–variate polynomial of degree d has $\binom{d+n}{n} = O(d^n)$ nonzero coefficients, we see that the dense or sparse representation of multivariate polynomials requires an exponential size, and their manipulation usually requires an exponential number of arithmetic operations with respect to the parameters d and n. In order to avoid this exponential behavior, we are going to use an alternative encoding of input and intermediate results of our computations by means of straight–line programs (cf. [9]). A *straight–line program* β in $\mathbb{K}(X) := \mathbb{K}(X_1, \ldots, X_n)$ is a finite sequence of rational functions $(F_1, \ldots, F_k) \in \mathbb{K}(X)^k$ such that for $1 \le i \le k$, the function F_i is either an element of the set $\{X_1, \ldots, X_n\}$, or an element of $\mathbb{K}$ (a *parameter*), or there exist $1 \le i_1, i_2 < i$ such that $F_i = F_{i_1} \circ_i F_{i_2}$ holds, where $\circ_i$ is one of the arithmetic operations $+, -, \times, \div$. The straight–line program β is called *division–free* if $\circ_i$ is different from $\div$ for $1 \le i \le k$. A natural measure of the complexity of β is its *time* (cf. [38]) which is the total number of arithmetic operations performed during the evaluation. We say that the straight–line program β *computes* or *represents* a subset S of $\mathbb{K}(X)$ if $S \subset \{F_1, \ldots, F_k\}$ holds.

2.2. The algorithmic model. Our algorithms are of *Monte Carlo* or *BPP* type (see, e.g., [2], [18]), i.e., they return the correct output with a probability of at least a fixed value strictly greater than 1/2. This in particular implies that the error probability can be made arbitrarily small by iteration of the algorithms. The probabilistic aspect of our algorithms is related to certain random choices of points with coordinates in a given finite field not annihilating certain nonzero polynomials. In order to perform a given random choice with a prescribed probability of success, we must know how many zeros the polynomial under consideration has. For this purpose, we have the following classical result, first shown by Oystein Ore in 1922.

PROPOSITION 2.1 ([32, Theorem 6.13]). *Let $\mathbb{K}$ be a finite field extension of $\mathbb{F}_q$ and let $F \in \overline{\mathbb{F}}_q[X]$ be a polynomial of degree d. The number of zeros of F in $\mathbb{K}$ is at most $d|\mathbb{K}|^{n-1}$.*

For the analysis of our algorithms, we shall interpret the statement of Proposition 2.1 in terms of probabilities. More precisely, assuming a uniform distribution of probability on the elements of the finite field $\mathbb{K}$, we have the following corollary, also known as the Zippel–Schwartz lemma in the computer algebra community (cf. [18, Lemma 6.44]).

COROLLARY 2.1. *Fix $\mu > 0$ and suppose that $|\mathbb{K}| > \mu d$ holds. Then the probability of choosing $x \in \mathbb{K}^n$ with $F(x) \neq 0$ is at least $1 - 1/\mu$.*

2.3. Cost of the basic operations. In this section we briefly review the cost of the basic algorithms we shall employ. The cost of such algorithms will be frequently stated in terms of the quantity

$$\mathsf{M}(m) := m \log^2 m \log\log m.$$

If $\mathbb{K}$ is a finite field, an arithmetic operation in $\mathbb{K}$ requires $O(\mathsf{M}(\log|\mathbb{K}|))$ bit operations. More generally, for a given domain R, the number of arithmetic operations in R necessary to compute the multiplication or division with remainder of two univariate polynomials in $R[T]$ of degree at most m is $O\big(\mathsf{M}(m)/\log(m)\big)$ (cf. [18], [7]).

If R is any field, then we shall use algorithms based on the Extended Euclidean Algorithm (EEA for short) in order to compute the gcd or the resultant of two univariate polynomials in $R[T]$ of degree at most m with $O\big(\mathsf{M}(m)\big)$ arithmetic operations in R (cf. [18],[7]).

Finally, we recall that for the cost $O(n^\omega)$ of the multiplication of two matrices of size $n \times n$ with coefficients in R, we have $\omega < 2.376$ (cf. [7]).

3. Geometric solutions. We shall use a representation of varieties which is well suited for algorithmic purposes (see, e.g., [20], [37], [21]). This representation is called a *geometric solution* or a *rational univariate representation* of the given variety. The notion of a geometric solution of an algebraic variety was first implicitly introduced in the works of Kronecker and König in the last years of the XIXth century. This section is devoted to motivate this notion and to describe certain underlying algorithmic aspects.

Let $\mathbb{K}$ be a perfect field and let $\overline{\mathbb{K}}$ denote its algebraic closure. We start with the definition of a ($\mathbb{K}$–definable) geometric solution of a zero-dimensional $\mathbb{K}$–variety. Let $V = \{P^{(1)}, \ldots, P^{(D)}\}$ be a zero–dimensional $\mathbb{K}$–variety of $\mathbb{A}^n$, and suppose that there exists a linear form $\mathcal{L} \in \mathbb{K}[X]$ which separates the points of V, i.e., which satisfies $\mathcal{L}(P^{(i)}) \neq \mathcal{L}(P^{(k)})$ if $i \neq k$. A geometric solution of V consists of

- a linear form $\mathcal{L} := \lambda \cdot X := \lambda_1 X_1 + \cdots + \lambda_n X_n \in \mathbb{K}[X]$ which separates the points of V,
- the minimal polynomial $m_\lambda := \prod_{1 \le i \le D}(T - \mathcal{L}(P^{(i)})) \in \mathbb{K}[T]$ of $\mathcal{L}$ in V (where T is a new variable),

- polynomials $w_1, \ldots, w_n \in \mathbb{K}[T]$ with $\deg w_j < D$ for every $1 \leq j \leq n$ satisfying the identity:

$$V = \{(w_1(\eta), \ldots, w_n(\eta)) \in \overline{\mathbb{K}}^n;\ \eta \in \overline{\mathbb{K}},\ m_\lambda(\eta) = 0\}.$$

Next, we define this notion for irreducible $\mathbb{K}$–varieties of dimension greater than 0. Let $V \subset \mathbb{A}^n$ be an irreducible $\mathbb{K}$-variety of dimension $r > 0$ and degree δ. Suppose that the indeterminates $X_1, \ldots, X_r$ form a separable transcendence basis of the field extension $\mathbb{K} \hookrightarrow \mathbb{K}(V)$, that is, $\mathbb{K}(X_1, \ldots, X_r) \hookrightarrow \mathbb{K}(V)$ is a finite, separable field extension. Denote by D its degree. In particular, we have that the linear projection $\pi : V \to \mathbb{A}^r$ defined by $\pi(x) := (x_1, \ldots, x_r)$ is a dominant morphism of degree D. From the behavior of the degree of a variety under linear maps (see, e.g., [22, Lemma 2]), it follows that $D \leq \delta$ holds.

Let $\Lambda := (\Lambda_1, \ldots, \Lambda_n)$ be a vector of new indeterminates. Observe that the extension $V_{\mathbb{K}(\Lambda)}$ of V to $\mathbb{A}^n(\overline{\mathbb{K}(\Lambda)})$ is an irreducible $\mathbb{K}(\Lambda)$–variety of dimension r and the coordinate ring of $V_{\mathbb{K}(\Lambda)}$ as a $\mathbb{K}(\Lambda)$–variety is isomorphic to $\mathbb{K}(\Lambda) \otimes_{\mathbb{K}} \mathbb{K}[V]$. Consider the generic linear form

$$\mathcal{L}_\Lambda := \Lambda \cdot X := \Lambda_1 X_1 + \cdots + \Lambda_n X_n. \tag{3.1}$$

Let $\xi_1, \ldots, \xi_n \in \mathbb{K}[V]$ be the coordinate functions induced by $X_1, \ldots, X_n$, and set $\widehat{\mathcal{L}} := \mathcal{L}_\Lambda(\xi_1, \ldots, \xi_n) \in \mathbb{K}(\Lambda) \otimes_{\mathbb{K}} \mathbb{K}[V]$. Since $\xi_1, \ldots, \xi_r, \widehat{\mathcal{L}}$ are algebraically dependent over $\mathbb{K}(\Lambda)$, there exists an irreducible polynomial $m_\Lambda \in \mathbb{K}[\Lambda, X_1, \ldots, X_r, T]$, separable with respect to T, such that the following identity holds in $\mathbb{K}(\Lambda) \otimes_{\mathbb{K}} \mathbb{K}[V]$:

$$m_\Lambda(\Lambda, \xi_1, \ldots, \xi_r, \widehat{\mathcal{L}}) = 0. \tag{3.2}$$

From, e.g., [39, Proposition 1], we deduce the following bounds:
- $\deg_T m_\Lambda = D$,
- $\deg_{X_1, \ldots, X_r} m_\Lambda \leq \delta$,
- $\deg_\Lambda m_\Lambda \leq \delta$.

Taking partial derivatives at both sides of equation (3.2) we deduce that for every $j = 1, \ldots, n$ the identity

$$\frac{\partial m_\Lambda}{\partial \Lambda_j}(\Lambda, \xi_1, \ldots, \xi_r, \widehat{\mathcal{L}}) + \xi_j \frac{\partial m_\Lambda}{\partial T}(\Lambda, \xi_1, \ldots, \xi_r, \widehat{\mathcal{L}}) = 0 \tag{3.3}$$

holds in $\mathbb{K}(\Lambda) \otimes_{\mathbb{K}} \mathbb{K}[V]$. As a consequence of the separability of m_Λ with respect to T we see that the polynomial $\partial m_\Lambda / \partial T$ is nonzero.

Assume that there exists $\lambda \in \mathbb{K}^n$ such that the linear form $\mathcal{L} := \lambda \cdot X$ induces a primitive element of the separable field extension $\mathbb{K}(X_1, \ldots, X_r) \hookrightarrow \mathbb{K}(V)$. Let ℓ be the coordinate function of $\mathbb{K}[V]$ defined by $\mathcal{L}$. From the fact that $\deg_T m_\Lambda = D$ holds, it follows that $m_\Lambda(\lambda, X_1, \ldots, X_r, T)$ is the minimal polynomial of ℓ in the field extension $\mathbb{K}(X_1, \ldots, X_r) \hookrightarrow \mathbb{K}(V)$.

Setting

$$m_\lambda(X_1,\ldots,X_r,T) := m_\Lambda(\lambda,X_1,\ldots,X_r,T),$$
$$v_j(X_1,\ldots,X_r,T) := -\frac{\partial m_\Lambda}{\partial \Lambda_j}(\lambda,X_1,\ldots,X_r,T),$$

and substituting λ for Λ in (3.2)–(3.3), we obtain the following identities of $\mathbb{K}[V]$:

$$\begin{aligned} m_\lambda(\xi_1,\ldots,\xi_r,\ell) &= 0,\\ \frac{\partial m_\lambda}{\partial T}(\xi_1,\ldots,\xi_r,\ell)\xi_j - v_j(\xi_1,\ldots,\xi_r,\ell) &= 0, \end{aligned} \tag{3.4}$$

which show that the polynomials

$$m_\lambda(X_1,\ldots,X_r,\mathcal{L}),\ \frac{\partial m_\lambda}{\partial T}(X_1,\ldots,X_r,\mathcal{L})X_j - v_j(X_1,\ldots,X_r,\mathcal{L})\ (1\le j\le n),$$

belong to $\mathbf{I}(V)$. We observe that the polynomials $v_1,\ldots,v_n$ have coefficients in $\mathbb{K}$ and satisfy the conditions $\deg v_j \le \delta$ and $\deg_T v_j \le D$.

Finally, we remark that the polynomials $m_\lambda(X_1,\ldots,X_r,\mathcal{L})$ and $(\partial m_\lambda/\partial T)(X_1,\ldots,X_r,\mathcal{L})X_j - v_j(X_1,\ldots,X_r,\mathcal{L})$ $(r+1\le j\le n)$ constitute a system of equations for the variety V in the Zariski dense open subset $V\cap\{(\partial m_\lambda/\partial T)(X,\mathcal{L})\neq 0\}$ of V. This motivates the definition of a geometric solution of an irreducible $\mathbb{K}$–variety of arbitrary dimension:

DEFINITION 3.1. *With assumptions as above, a* geometric solution *of V consists of the following items:*

- *a linear form $\mathcal{L} := \lambda\cdot X \in \mathbb{K}[X]$ which induces a primitive element ℓ of the field extension $\mathbb{K}(X_1,\ldots,X_r)\hookrightarrow\mathbb{K}(V)$,*
- *the minimal polynomial $m_\lambda \in \mathbb{K}[X_1,\ldots,X_r][T]$ of ℓ,*
- *a generic* parametrization *of the variety V by the zeros of m_λ, of the form $(\partial m_\lambda/\partial T)X_j - v_j$ $(r+1\le j\le n)$, with $v_j\in\mathbb{K}[X_1,\ldots,X_r][T]$, $\deg_T v_j < D$, $\deg_X v_j \le \delta$ and $(\partial m_\lambda/\partial T)(\mathcal{L})X_j - v_j(\mathcal{L}) \in \mathbf{I}(V)$.*

We observe that the polynomial $m_\lambda \in \mathbb{K}[X_1,\ldots,X_r][T]$ of the second item of the previous definition can be also defined as follows: consider the linear map $\pi_\lambda : V \to \mathbb{A}^{r+1}$ defined by $\pi_\lambda(x) := (x_1,\ldots,x_r,\lambda\cdot x)$. The Zariski closure of $\pi_\lambda(V)$ is a $\mathbb{K}$–hypersurface H of $\mathbb{A}^{r+1}$ of degree at most δ, which is indeed defined by $m_\lambda(X_1,\ldots,X_r,T) = 0$.

3.1. Algorithmic aspects of the computation of a geometric solution. From the algorithmic point of view, the crucial step towards the computation of a geometric solution of a variety V consists in the computation of the minimal polynomial m_Λ of the generic linear form $\mathcal{L}_\Lambda$. In this section we shall show how we can derive an algorithm for computing a geometric solution of an r–dimensional $\mathbb{K}$–variety V from a procedure for computing the minimal polynomial of the generic linear form $\mathcal{L}_\Lambda$ (cf. [1], [21], [39]).

Assume that we have already chosen $\lambda \in \mathbb{K}^n$ such that the linear form $\mathcal{L} := \lambda \cdot X$ induces a primitive element of the separable field extension $\mathbb{K}(X_1, \ldots, X_r) \hookrightarrow \mathbb{K}(V)$. Let $m_\lambda \in \mathbb{K}[X_1, \ldots, X_r, T]$ be its minimal polynomial.

Suppose that we are given an algorithm Φ over $\mathbb{K}(\Lambda)$ for computing the minimal polynomial of the linear form $\mathcal{L}_\Lambda = \Lambda \cdot X$. Suppose further that the vector $(\lambda_1, \ldots, \lambda_n)$ of coefficients of $\mathcal{L}$ does not annihilate any denominator in $\mathbb{K}[\Lambda]$ of any intermediate result of the algorithm Φ. In order to compute the polynomials $v_{r+1}, \ldots, v_n$ of Definition 3.1, we observe that the Taylor expansion of $m_\Lambda(\Lambda, X_1, \ldots, X_r, T)$ in powers of $\Lambda - \lambda := (\Lambda_1 - \lambda_1, \ldots, \Lambda_n - \lambda_n)$ has the following expression:

$$m_\Lambda(\Lambda, X_1, \ldots, X_r, T) = m_\lambda(X_1, \ldots, X_r, T) +$$
$$\sum_{j=1}^{n} \Big(\frac{\partial m_\lambda}{\partial T}(X_1, \ldots, X_r, T) X_j - v_j(X_1, \ldots, X_r, T)\Big)(\Lambda_j - \lambda_j) \mod (\Lambda - \lambda)^2.$$

We shall compute this first–order Taylor expansion by computing the first–order Taylor expansion of each intermediate result in the algorithm Φ. In this way, each arithmetic operation in $\mathbb{K}(\Lambda)$ arising in the algorithm Φ becomes an arithmetic operation between two polynomials of $\mathbb{K}[\Lambda]$ of degree at most 1, and is truncated up to order $(\Lambda - \lambda)^2$. Since the first–order Taylor expansion of an addition, multiplication or division of two polynomials of $\mathbb{K}[\Lambda]$ of degree at most 1 requires $O(n)$ arithmetic operations in $\mathbb{K}$, we have that the whole step requires $O(n\mathsf{T})$ arithmetic operations in $\mathbb{K}$, where T is the number of arithmetic operations in $\mathbb{K}(\Lambda)$ performed by the algorithm Φ. Summarizing, we have the following result:

LEMMA 3.1. *Suppose that we are given:*

1. *an algorithm Φ in $\mathbb{K}(\Lambda)$ which computes the minimal polynomial $m_\Lambda \in \mathbb{K}[\Lambda, X_1, \ldots, X_r, T]$ of $\mathcal{L}_\Lambda := \Lambda \cdot X$ with T arithmetic operations in $\mathbb{K}(\Lambda)$,*
2. *a separating linear form $\mathcal{L} := \lambda \cdot X \in \mathbb{K}[X]$ such that the vector λ does not annihilate any denominator in $\mathbb{K}[\Lambda]$ of any intermediate result of the algorithm Φ.*

Then a geometric solution of the variety V can be (deterministically) computed with $O\big(n(\mathsf{T} + \mathsf{M}(D))\big)$ arithmetic operations in $\mathbb{K}$.

4. Preparation of the input data. Let $F_1, \ldots, F_n \in \mathbb{F}_q(X)$ be rational functions having a reduced representation $F_i = P_i/Q_i$ with numerator and denominator of degree at most d for $1 \leq i \leq n$. Consider the rational map $F : \mathbb{A}^n \to \mathbb{A}^n$ defined by $F(x) := (F_1(x), \ldots, F_n(x))$. Since the rational functions $F_1, \ldots, F_n$ have coefficients in $\mathbb{F}_q$, we see that the restriction of F to $\mathbb{F}_q^n$ induces a (partially–defined) mapping from $\mathbb{F}_q^n$ to $\mathbb{F}_q^n$ which we shall also denote by F, with a slight abuse of notation.

4.1. The graph of the mapping F. Let $Y := (Y_1, \ldots, Y_n)$ be a vector of new indeterminates. Our algorithm shall deal with the $\mathbb{F}_q$–variety

$V \subset \mathbb{A}^{2n}$ representing the Zariski closure of the graph of the mapping F. More precisely, let $F_i := P_i/Q_i$ be a reduced fraction representing the rational function F_i for $1 \le i \le n$ and set $Q := Q_1 \cdots Q_n$. Let $I \subset \mathbb{F}_q[X, Y]$ be the ideal generated by the polynomials $Q_i(X) - Y_i P_i(X)$ $(1 \le i \le n)$. Then we define V as

$$V := V(I : Q^\infty),$$

where Q^∞ denotes the multiplicatively closed subset of $\mathbb{F}_q[X, Y]$ generated by 1 and Q and $(I : Q^\infty)$ denotes the saturation of the ideal I by Q^∞, that is, $(I : Q^\infty) := \{P \in \mathbb{F}_q[X, Y]; PQ^s \in I \text{ for some } s \in \mathbb{Z}_{\ge 0}\}$.

Let $\pi : V \to \mathbb{A}^n$ be the projection mapping defined by $\pi(x, y) := y$. In what follows, we shall assume that F and π satisfy the following conditions, which are usually met in the cryptographic situations we are interested in:

(*i*) F is partially defined over $\mathbb{F}_q^n$.

(*ii*) π is a dominant mapping. In particular, the fiber $V_y := \pi^{-1}(y)$ is a zero–dimensional subvariety of V for a generic $y \in \mathbb{A}^n$.

We observe that (*i*) is required by most cryptographic schemes based on multivariate equations (see, e.g., [30, Chapter 4], [43]), while (*ii*) is required for example in cryptographic schemes based on "tractable" rational maps (see, e.g., [42], [43, Section 6]).

Assumption (*ii*) and the definition of V imply that V is an absolutely irreducible $\mathbb{F}_q$–variety of dimension n. Indeed, it is easy to see that $\mathbb{F}_q(V)$ is isomorphic to $\mathbb{F}_q(X)$, which implies that V is absolutely irreducible and of dimension n. Further consequences of our assumptions are that the set of variables Y is algebraically independent in $\mathbb{F}_q(V)$ and the polynomials $Q_i(X)Y_i - P_i(X)$ $(1 \le i \le n)$ generate a radical ideal of the localization $\mathbb{F}_q[X, Y]_{Q^\infty}$.

In the sequel, we shall denote by δ the degree of V and by D the degree of the morphism $\pi : V \to \mathbb{A}^n$.

4.2. Random choices. Let $\mathcal{L} := \lambda \cdot X \in \overline{\mathbb{F}}_q[X]$ be a linear form such that the corresponding coordinate function of $\overline{\mathbb{F}}_q[V]$ is a primitive element of the field extension $\overline{\mathbb{F}}_q(Y) \hookrightarrow \overline{\mathbb{F}}_q(V)$. In particular, the minimal polynomial $m_\lambda \in \overline{\mathbb{F}}_q[Y, T]$ of the coordinate function defined by $\mathcal{L}$ satisfies the degree estimate $\deg_T m_\lambda = D$. By the remark after Definition 3.1 we see that V is birationally equivalent to the hypersurface $H \subset \mathbb{A}^{n+1}$ defined by the polynomial $m_\lambda \in \overline{\mathbb{F}}_q[Y, T]$. We observe that the fact that V is absolutely irreducible implies that H, and thus m_λ, is absolutely irreducible.

For a given point $y^{(0)} \in \mathbb{F}_q^n$, we denote by $V_{y^{(0)}}$ the π–fiber of $y^{(0)}$. In order to compute the points of the set $F^{-1}(y^{(0)}) \cap \mathbb{F}_q^n$, or equivalently, the set $V_{y^{(0)}} \cap \mathbb{F}_q^{2n}$, we shall deform the system $F(X) = y^{(0)}$ into a system $F(X) = F(x^{(1)})$ with a point $x^{(1)}$ randomly chosen in a suitable finite field extension $\mathbb{K}$ of $\mathbb{F}_q$ to be determined. The kind of deformations we shall apply is inspired by the approach of [34]. In our next result we establish suitable bounds on the degree of the genericity conditions underlying the choice of $x^{(1)}$.

LEMMA 4.1. *There exists a nonzero polynomial $A \in \overline{\mathbb{F}}_q[X]$ of degree at most $3d\delta^4$ such that for any $x \in \mathbb{A}^n$ with $A(x) \neq 0$, the point $y := F(x)$ satisfies the following conditions:*

(i) y is a lifting point of π,

(ii) Let S be a new indeterminate and let $I_C \subset \overline{\mathbb{F}}_q[S, X]$ be the ideal

$$I_C := \Big(P_i(X) - Q_i(X)\big(y + (S-1)(y - y^{(0)})\big); 1 \leq i \leq n\Big).$$

Then the curve $C := V(I_C : Q^\infty) \subset \mathbb{A}^{n+1}$ is absolutely irreducible.

Proof. Let $m_\lambda \in \overline{\mathbb{F}}_q[Y][T]$ be the minimal (primitive) polynomial of the linear form $\mathcal{L} := \lambda \cdot X \in \overline{\mathbb{F}}_q[X]$. Let $A_1^* \in \overline{\mathbb{F}}_q[Y]$ denote the discriminant of m_λ with respect to the variable T. From the absolutely irreducibility of m_λ we conclude that $A_1^* \neq 0$ holds. Furthermore, for any $y \in \mathbb{A}^n$ with $A_1^*(y) \neq 0$ we have that the fiber V_y consists of D distinct points, that is, y is a lifting point of π. Hence, the nonvanishing of the polynomial $A_1^* \in \overline{\mathbb{F}}_q[Y]$ represents a suitable genericity condition underlying the choice of a lifting point $y \in \mathbb{A}^n$.

In order to obtain a genericity condition underlying the choice of a point $x \in \mathbb{A}^n$ for which $y := \pi(x)$ is a lifting point, consider a reduced representation $A_1^*(F(X)) = P_1^*/Q_1^*$ of the rational function defined by $A_1^*(F(X))$ and set $A_1 := P_1^* Q_1^*$. By definition it follows that $A_1 \in \overline{\mathbb{F}}_q[X]$ has degree bounded by $(2D-1)d\delta$. Since F is a dominant mapping, we have that there exists $x \in \mathbb{A}^n$ such that $A_1(x) \neq 0$ holds (see, e.g., [40, II.6.3, Theorem 4]). This implies that A_1 is a nonzero polynomial.

Next we consider a reduced representation

$$m_\lambda\big(F(X) + (S-1)(F(X) - y^{(0)}), T\big) = \frac{\mathcal{P}_1(X)\mathcal{P}_2(X, S, T)}{\mathcal{Q}(X)}$$

of the rational function $m_\lambda\big(F(X) + (S-1)(F(X) - y^{(0)}), T\big) \in \overline{\mathbb{F}}_q(X)[S, T]$, where $\mathcal{P}_2(X, S, T)$ is a primitive polynomial of $\overline{\mathbb{F}}_q[X][S, T]$. Observe that such a representation is unique up to scaling by nonzero elements of $\overline{\mathbb{F}}_q$. Set

$$\widetilde{m}_\lambda(X, S, T) := \frac{\mathcal{Q}(X)}{\mathcal{P}_1(X)} m_\lambda\big(F(X) + (S-1)(F(X) - y^{(0)}), T\big) = \mathcal{P}_2(X, S, T).$$

Let $x \in \mathbb{A}^n$ be any point with $Q(x) \neq 0$. Then the value $F(x)$ is well–defined, and hence $m_\lambda(F(x) + (S-1)(F(x) - y^{(0)}), T)$ and $\widetilde{m}_\lambda(x, S, T)$ are both well–defined nonzero polynomials of $\overline{\mathbb{F}}_q[S, T]$ of degree D. As a consequence, for any lifting point y of π and any $x \in V_y$, the polynomial $\widetilde{m}_\lambda(x, 1, T)$ is a nonzero scalar multiple of $m_\lambda(y, T)$, and thus a separable element of $\overline{\mathbb{F}}_q[T]$ of degree D.

Following [28, Theorem 5], in the version of [10, Theorem 3.6], there exists a polynomial $A_2^* \in \overline{\mathbb{F}}_q[Y]$ of degree bounded by $2\delta^4 + \delta$ such that for any $y \in \mathbb{A}^n$ with $A_2^*(y) \neq 0$ the polynomial $m_\lambda(y + (S-1)(y - y^{(0)}), T)$ is absolutely irreducible. Let $A_2 \in \overline{\mathbb{F}}_q[X]$ be the numerator of a reduced representation of the rational function $A_2^*(F(X)) \in \overline{\mathbb{F}}_q(X)$. It follows that A_2 has degree bounded by $2d\delta^4 + d\delta$ and, for any $x \in \mathbb{A}^n$ with $A_2(x) \neq 0$, the polynomial $\widetilde{m}_\lambda(x, S, T)$ is absolutely irreducible.

Let $A := A_1 A_2$. Observe that $A \in \overline{\mathbb{F}}_q[X]$ and has degree at most $3d\delta^4$. Now, if we consider any point $x \in \mathbb{A}^n$ satisfying $A(x) \neq 0$ and set $y := F(x)$, we claim that conditions (i) and (ii) of the statement of the lemma are satisfied. Indeed, $A_1(x) \neq 0$ implies that $A_1^*(y) \neq 0$, that is, the discriminant of $m_\lambda(y, T)$ with respect to T is nonzero. We deduce that $m_\lambda(y, T)$ has D distinct roots and therefore, y is a lifting point of π. Finally, since y is a lifting point of π and $A_2(x) \neq 0$, the polynomial $\widetilde{m}_\lambda(x, S, T)$ is absolutely irreducible and hence, so is the curve $\mathcal{C}$. □

We remark that in the case of a field of large characteristic, say $char(\mathbb{F}_q) \geq 2\delta^2$ or $char(\mathbb{F}_q) \geq \delta(\delta - 1) + 1$, the bound of the statement of Lemma 4.1 can be improved applying the approach of [17] or [31] respectively. More precisely, applying [31, Theorem 6] (see also [17, Theorem 5.1] for a slightly worse bound) it follows that there exists a nonzero polynomial $A \in \overline{\mathbb{F}}_q[X]$ of degree at most $4d\delta^2$ for which the conditions of the statement of Lemma 4.1 hold. Nevertheless, taking into account that in cryptographic applications fields of characteristic 2 are very common, we shall not pursue the subject any further.

Suppose that we have already chosen a point $x \in \mathbb{A}^n$ satisfying the conditions of Lemma 4.1 and let $y := F(x)$. Let $\Lambda := (\Lambda_1, \ldots, \Lambda_n)$ be a vector of new indeterminates.

LEMMA 4.2. *There exists a nonzero polynomial $B \in \overline{\mathbb{F}}_q[\Lambda]$ of degree at most $2D^2$ such that for any $\lambda \in \mathbb{A}^n$ with $B(\lambda) \neq 0$, the linear form $\mathcal{L} := \lambda \cdot X$ separates the points of V_y and $V_{y^{(0)}}$.*

Proof. Let $V_y \cup V_{y^{(0)}} := \{P_1, \ldots, P_{D'}\}$. We consider the generic linear form $\mathcal{L}_\Lambda := \Lambda \cdot X$ and define

$$B(\Lambda) := \prod_{1 \leq i < j \leq D'} (\mathcal{L}_\Lambda(P_i) - \mathcal{L}_\Lambda(P_j)).$$

Since $D' \leq 2D$ holds, it follows that $B \in \overline{\mathbb{F}}_q[\Lambda]$ is a nonzero polynomial of degree at most $2D^2$. Any $\lambda \in \mathbb{A}^n$ not annihilating B provides a linear form $\mathcal{L}$ that separates the points of V_y and $V_{y^{(0)}}$. □

Now we can determine the degree of a field extension $\mathbb{K}$ of $\mathbb{F}_q$ for which the existence of points $\lambda, x \in \mathbb{K}^n$ satisfying Lemmas 4.1 and 4.2 can be assured. Our next result states that for a random choice of the coordinates of λ and x in a field extension $\mathbb{K}$ of $\mathbb{F}_q$ of suitable degree the statements of Lemmas 4.1 and 4.2 hold with high probability of success.

COROLLARY 4.1. *With notations as in Lemmas 4.1 and 4.2, fix $\mu > 0$ and let $\mathbb{K}$ be a finite field extension of $\mathbb{F}_q$ such that $|\mathbb{K}| > 4\mu d\delta^4$ holds. Then a random choice of (λ, x) in $\mathbb{K}^{2n}$ satisfies the condition $(AB)(\lambda, x) \neq 0$ with error probability at most $1/\mu$.*

Proof. By Proposition 2.1, the number of zeros in $\mathbb{K}^n$ of the polynomial A is at most $3d\delta^4|\mathbb{K}|^{n-1}$. Then a random choice of $x \in \mathbb{K}^n$ satisfies $A(x) \neq 0$ with probability at least $1 - 3d\delta^4/|\mathbb{K}| \geq 1 - 3/4\mu$. Given such a choice, a random choice of $\lambda \in \mathbb{K}^n$ satisfies $B(\lambda) \neq 0$ with probability at least $1 - 2D^2/|\mathbb{K}| \geq 1 - 1/4\mu$. This shows that a random choice $(\lambda, x) \in \mathbb{K}^{2n}$ satisfies $(AB)(\lambda, x) \neq 0$ with probability at least $(1 - 3/4\mu)(1 - 1/4\mu) \geq 1 - 1/\mu$. □

We remark that the polynomial AB of statement of Corollary 4.1 will not be computed during the execution of our algorithm, and therefore our algorithm will proceed with a random choice $(\lambda, x) \in \mathbb{K}$ for which the identity $AB(\lambda, x) = 0$ might hold. In such an unlikely event, certain intermediate values which are expected to be nonzero are equal to zero, and the algorithm must be restarted with another random choice of (λ, x).

A second remark is that, as we do not know in general the values of D and δ *a priori* (although in some cryptosystems such values are known), in order to determine the size of the field $\mathbb{K}$ these values can be estimated by d^n. This will not increase significatively the cost of our algorithm, since the cost depends linearly on the logarithm of $|\mathbb{K}|$.

5. The algorithm. Let $\mathbb{K}$ be a finite field extension of $\mathbb{F}_q$ whose cardinality will be determined later. Let $(\lambda, x^{(1)}) \in \mathbb{K}^{2n}$ be a point randomly chosen. By Corollary 4.1 we have that $(\lambda, x^{(1)})$ satisfies the conditions in the statements of Lemmas 4.1 and 4.2 with error probability at most $4d\delta^4/|\mathbb{K}|$. This means that with such an error probability the following assertions hold:

- $y^{(1)} := F(x^{(1)})$ is a well-defined lifting point of π;
- let $I_{\mathcal{C}} := (P_i(X) - Q_i(X)(y_i^{(1)} + (S-1)(y_i^{(1)} - y_i^{(0)}))$. Then
$$\mathcal{C} := V(I_{\mathcal{C}} : Q^\infty) \tag{5.1}$$
is an absolutely irreducible curve of $\mathbb{A}^{n+1}$;
- the linear form $\mathcal{L} := \lambda \cdot X \in \mathbb{K}[X]$ separates the points of the fibers $V_{y^{(1)}}$ and $V_{y^{(0)}}$.

In what follows, we shall assume that all these conditions hold.

We consider the projection $\pi_S : \mathcal{C} \to \mathbb{A}^1$ defined by $\pi_S(s, x) := s$. We have that π_S is a dominant mapping of degree D, that $S = 1$ is a lifting point of π_S and that the identities $\pi_S^{-1}(1) = \{1\} \times \mathcal{C}_1$ and $\pi_S^{-1}(0) = \{0\} \times \mathcal{C}_0$ hold, where $\mathcal{C}_1 := F^{-1}(y^{(1)})$ and $\mathcal{C}_0 := F^{-1}(y^{(0)})$ denote the fibers defined by $y^{(1)}$ and $y^{(0)}$ respectively. Since $\mathcal{L}$ separates the points of $V_{y^{(1)}}$ it follows that $\mathcal{L}$ is a primitive element of the field extension $\mathbb{K}(S) \hookrightarrow \mathbb{K}(\mathcal{C})$.

The algorithm that computes all the points of $V_{y^{(0)}}$ may be divided in three main parts, which will be considered in Sections 5.1, 5.2 and 5.3 below. In the first step, we compute the minimal primitive polynomial $m_S(S, T)$ of $\mathcal{L}$ in the field extension $\mathbb{K}(S) \hookrightarrow \mathbb{K}(\mathcal{C})$. For this purpose, we apply a Newton–Hensel iteration to the rational point $x^{(1)}$ in order to obtain the vector of power series $\Psi \in \mathbb{K}[\![S-1]\!]^n$ which parametrizes the branch of C passing through $(x^{(1)}, y^{(1)})$, truncated up to a suitable precision. It turns out that the least-degree nonzero polynomial $m_S(S,T) \in \mathbb{K}[S,T]$ which annihilates the power series $\mathcal{L}(\Psi)$ up to a certain precision equals the minimal polynomial of the coordinate function defined by $\mathcal{L}$ in the field extension $\mathbb{K}(S) \hookrightarrow \mathbb{K}(C)$ (see Lemma 5.1 below).

In the second step we extend the computation of the minimal polynomial $m_S(S,T)$ of $\mathcal{L}$ in the field extension $\mathbb{K}(S) \hookrightarrow \mathbb{K}(C)$ to the computation of a geometric solution of the curve C, applying the algorithm underlying Lemma 3.1. Finally, in the third step we find the coordinates of the q–rational points of $V_{y^{(0)}}$. In order to do this, we first obtain a geometric solution $m_S(0,T), w_1(T), \ldots, w_n(T)$ of the zero–dimensional variety $C_0 = F^{-1}(y^{(0)})$, substituting 0 for S in the polynomials which form the geometric solution of C computed in the previous step. Then we easily obtain the q–rational points of C_0 among the points $x := (x_1, \ldots, x_n) \in \mathbb{A}^n$ satisfying the following equations:

$$m_S(0,T) = 0,\ T^{|\mathbb{K}|} - T = 0,\ x_i = w_i(T) \quad (1 \le i \le n).$$

The whole algorithm for computing the q–rational points of $F^{-1}(y^{(0)})$ may be briefly sketched as follows:

Algorithm 5.1.

1. *Choose the coefficients of a vector* $(\lambda, x^{(1)}) \in \mathbb{K}^{2n}$ *at random.*
2. *Set* $G(S,X) := F(X) - y^{(1)} - (S-1)(y^{(1)} - y^{(0)})$. *Compute the Newton–Hensel operator* $N_G(X) := X - J_F^{-1}(X)G(S,X)$.
3. *Compute* $\kappa := \lceil \log_2(2D\delta + 1) \rceil$ *iterations of the Newton–Hensel iterator* N_G *applied to* $x^{(1)}$. *Let* Ψ_κ *be the resulting vector of power series truncated up to order* $2D\delta + 1$.
4. *Find the least–degree nonzero polynomial* $m_S \in \mathbb{K}[S,T]$ *such that* $m_S(S, \mathcal{L}(\Psi_\kappa)) \equiv 0 \bmod (S-1)^{2D\delta+1}$ *holds. This is the minimal polynomial of* $\mathcal{L}$ *in* $\mathbb{K}(S) \hookrightarrow \mathbb{K}(C)$ *(see Lemma 5.1).*
5. *Obtain a geometric solution of the curve* C *applying the proof of Lemma 3.1.*
6. *Substitute 0 for* S *in the polynomials which form the geometric solution of* C *computed in the previous step. The univariate polynomials obtained form a complete description of* C_0 *(eventually including multiplicities).*
7. *Clean multiplicities of the polynomials computed in the previous step to obtain a geometric solution* $m_0, w_1, \ldots, w_n \in \mathbb{K}[T]$ *of the variety* C_0.
8. *Compute* $h := \gcd(m_0, T^{|\mathbb{K}|} - T)$ *and the roots* $\alpha^{(1)}, \ldots, \alpha^{(M)}$ *of* h *in* $\mathbb{K}$.
9. *Compute the* q*–rational points of* $C_0 = F^{-1}(y^{(0)})$ *as the intersection* $\{(w_1(\alpha^{(i)}), \ldots, w_n(\alpha^{(i)})); 1 \le i \le M\} \cap \mathbb{F}_q^n$.

We observe that, in order to determine the size of the field $\mathbb{K}$ and to execute steps (3)–(4), the values D and δ are required. Although in some cases these values are known *a priori*, we cannot in general assume that they are given. Concerning the determination of the field $\mathbb{K}$, from the complexity point of view we may simply estimate D and δ by d^n and proceed, since the cost of our procedure depends quasi-linearly on the value $\log_2(D\delta)$. On the other hand, for the execution of steps (3)–(5), the value $N := 2D\delta$ can be found by a process which, roughly speaking, starts with the value $N = 2$, and incrementally doubles the value N until the output of steps (3)–(5) is a geometric solution of the curve $\mathcal{C}$. The efficiency of this process relies on the fact that one can efficiently check if a candidate to be a geometric solution of a given irreducible variety is actually a geometric solution. Such a modification would only contribute with logarithmic factors to the asymptotic cost of our procedure.

5.1. The computation of the polynomial m_S. We consider the factorization of $m_S(S,T)$ in the ring $\mathbb{K}[\![S-1]\!][T]$, where $\mathbb{K}[\![S-1]\!]$ denotes the power series ring in $S-1$. From the fact that $m_S(1,T)$ is separable of degree D we conclude that there exist D distinct power series $\sigma^{(1)},\dots,\sigma^{(D)} \in \mathbb{K}[\![S-1]\!]$ such that the monic version $\widetilde{m}_S$ of $m_S(S,T)$ can be factored as $\widetilde{m}_S = \prod_{i=1}^{D}(T-\sigma^{(i)})$. Furthermore, $\widetilde{m}_S(1,T)$ can be factored as $\widetilde{m}_S(1,T) = \prod_{i=1}^{D}(T-\sigma^{(i)}(1))$, where $\sigma^{(i)}(1)$ represents the constant term of $\sigma^{(i)}$ for $1 \le i \le D$.

Since $\widetilde{m}_S(1,T)$ is the minimal polynomial of the linear form $\mathcal{L}$ in the $\mathbb{K}$–algebra extension $\mathbb{K} \hookrightarrow \mathbb{K}[V_{y^{(1)}}]$, if we write $V_{y^{(1)}} := \{P_1,\dots,P_D\}$ we have that $\widetilde{m}_S(1,T) = \prod_{i=1}^{D}(T-\mathcal{L}(P_i))$. Given that $(x^{(1)},y^{(1)})$ belongs to the fiber $V_{y^{(1)}}$, there exists a power series $\sigma^{(i)}$ such that $\mathcal{L}(x^{(1)}) = \sigma^{(i)}(1)$. In order to simplify notations, we shall simply write σ instead of $\sigma^{(i)}$.

The algorithm that computes the polynomial $m_S(S,T)$ starts computing the power series σ truncated up to order $N+1$, where $N := 2D\delta$. Let $\sigma_N \in \overline{\mathbb{F}}_q[S]$ be the polynomial of degree at most N congruent to σ modulo $(S-1)^{N+1}$. Our next result shows that the polynomial $m_S(S,T)$ we want to compute can be obtained as the solution of a suitable congruence equation involving σ_N.

LEMMA 5.1. *Let $g \in \mathbb{K}[S,T]$ be a polynomial with $\deg_S g \le \delta$ and $\deg_T g \le D$ satisfying the following congruence*

$$g(S,\sigma_N) \equiv 0 \mod (S-1)^{N+1}. \tag{5.2}$$

Then m_S divides g in $\mathbb{K}[S,T]$.

Proof. Let $g \in \mathbb{K}[S,T]$ be a solution of (5.2) satisfying the conditions on the degree of the statement of the lemma. The resultant $h \in \mathbb{K}[S]$ of g and m_S with respect to T has degree at most N and belongs to the ideal generated by m_S and g. Since $m_S(S,\sigma_N)$ and $g(S,\sigma_N)$ are congruent to 0 modulo $(S-1)^{N+1}$ by hypothesis, we deduce that $h(S) \equiv 0 \mod (S-1)^{N+1}$ holds. Therefore, the fact that $\deg h \le N$ and $h(S) \equiv 0 \mod (S-1)^{N+1}$ holds imply $h = 0$. In particular, we derive the existence of a common factor of m_S and g in $\mathbb{K}(S)[T]$. Finally, taking into account the irreducibility of m_S in $\mathbb{K}(S)[T]$ and the Gauss lemma, we easily deduce the statement of the lemma. □

From Lemma 5.1 we conclude that m_S can be characterized as the nonzero solution of (5.2) of minimal (total) degree.

In order to find the least-degree nonzero solution of (5.2), we shall interpret (5.2) as a problem of Hermite–Padé approximation. Indeed, finding a nonzero solution of the congruence equation (5.2) is equivalent to finding $g_0, \ldots, g_D \in \mathbb{K}[S]$ with $\deg g_j \le \delta$ for $0 \le j \le D$ such that the following congruence equation holds:

$$g_0(S) + g_1(S)\sigma_N + \cdots + g_D(S)\sigma_N^D \equiv 0 \mod (S-1)^{N+1}. \tag{5.3}$$

We shall solve (5.3) applying an algorithm due to [8], which is based on fast linear–algebra algorithms for matrices of fixed displacement rank (cf. [7], [33]). This requires the computation of the successive powers $\sigma_N, \ldots, \sigma_N^D$ of the power series σ truncated up to order $N+1$.

The computation of σ_N is based on a multivariate Newton iteration over the power series ring $\mathbb{K}[\![S-1]\!]$, which we now describe. Substituting 1 for S in the polynomials defining the ideal I_C associated to the curve C of (5.1), we obtain the system $y^{(1)} = F(X)$. Since $y^{(1)}$ is a lifting point of π, it follows that none of the points of $V_{y^{(1)}}$ annihilate the denominator $Q_i(X)$ of the rational function F_i for $1 \le i \le n$. Furthermore, from, e.g., [10, Lemma 2.1] we conclude that none of the points of $V_{y^{(1)}}$ annihilate the determinant of the Jacobian matrix $J_F := (\partial F_i)/(\partial X_j)_{1 \le i,j \le n}$. In particular, $\det J_F(x^{(1)}) \ne 0$ holds.

Observe that the curve C of (5.1) is locally defined in a neighborhood of each point of $V_{y^{(1)}}$ by the equations $F_i(X) = y_i^{(1)} + (S-1)(y_i^{(1)} - y_i^{(0)})$ $(1 \le i \le n)$. Therefore, in order to compute the truncated power series σ_N we consider the Newton–Hensel operator N_G associated to the vector of rational functions $G(S,X) := F(X) - y^{(1)} - (S-1)(y^{(1)} - y^{(0)})$, namely,

$$N_G(X) := X - J_F^{-1}(X)G(S,X).$$

Let $N_G^{(k)}$ denote the k–fold iteration of N_G and define $\Psi_k := N_G^{(k)}(x^{(1)}) \in \mathbb{K}[\![S-1]\!]^n$ for $k \ge 0$. Then it is well known that the following congruence relation holds:

$$G(S, \Psi_k) \equiv 0 \mod (S-1)^{2^k}. \tag{5.4}$$

Since $Q_i(\Psi_k)(1) \ne 0$ holds for $1 \le i \le n$, from (5.4) we deduce that

$$P_i(\Psi_k) \equiv Q_i(\Psi_k)(y_i^{(1)} + (S-1)(y_i^{(1)} - y_i^{(0)})) \mod (S-1)^{2^k}. \tag{5.5}$$

Since the polynomial $m_S(S, \mathcal{L}(X))$ belongs to the ideal of $\mathbb{K}[S,X]_{Q^\infty}$ generated by $P_i(X) - Q_i(X)(y_i^{(1)} + (S-1)(y_i^{(1)} - y_i^{(0)}))$ $(1 \le i \le n)$, from (5.5) we conclude that

$$m_S(S, \mathcal{L}(\Psi_k)) \equiv 0 \mod (S-1)^{2^k}.$$

From the identity $\mathcal{L}(\Psi_k)(1) = \mathcal{L}(x^{(1)})$ we deduce that $\mathcal{L}(\Psi_k) \equiv \sigma \bmod (S-1)^{2^k}$ holds. Hence, we obtain σ_N as the power series $\mathcal{L}(\Psi_\kappa)$ with $\kappa := \lceil \log_2(N+1) \rceil$ truncated up to order $N+1$. From σ_N we easily compute the powers $\sigma_N^2, \ldots, \sigma_N^D$ by successive multiplication and truncation.

We may summarize the algorithm underlying the above considerations as follows:

ALGORITHM 5.2 (Computation of the powers of σ_N).

1. Set $\kappa := \lceil \log_2(N+1) \rceil$ and $\Psi_0 := x^{(1)}$.

2. Compute $\Psi_{k+1} := N_G(\Psi_k) \bmod (S-1)^{2^{k+1}}$ for $0 \leq k \leq \kappa - 1$.

3. Compute $\sigma_N := \mathcal{L}(\Psi_\kappa) \bmod (S-1)^{N+1}$.

4. Compute $\sigma_N^{j+1} := \sigma_N \cdot \sigma_N^j \bmod (S-1)^{N+1}$ for $1 \leq j \leq D-1$.

The following proposition provides a complexity estimate of the procedure just described:

PROPOSITION 5.1. *If the rational functions $F_1, \ldots, F_n$ are evaluated with T operations in $\mathbb{F}_q$, the powers $\sigma_N, \ldots, \sigma_N^D$, truncated up to order $N+1$, can be deterministically computed with $O\big((\mathsf{T} + n^{1+\omega})\mathsf{M}(D\delta)\big)$ arithmetic operations in $\mathbb{K}$.*

Proof. First, from the Baur–Strassen theorem [5] it follows that the entries of J_F can be computed with $O(\mathsf{T})$ arithmetic operations. Then, the determinant and adjoint matrix of J_F can be evaluated with $O(\mathsf{T} + n^{1+\omega})$ arithmetic operations (see, e.g., [7]).

In order to compute the $(k+1)$th iteration $\Psi_{k+1} := N_G(\Psi_k)$ from Ψ_k, we compute the inverse matrix $J_F^{-1}(\Psi_k)$ as the multiplication $J_F^{-1}(\Psi_k) = \det J_F(\Psi_k)^{-1} \cdot Adj\big(J_F(\Psi_k)\big)$ of the reciprocal of the (truncated) power series $\det J_F(\Psi_k)$ by each entry of the adjoint matrix $Adj\big(J_F(\Psi_k)\big)$. Using fast power series inversion we can compute $\det J_F(\Psi_k)^{-1}$ with $O\big((\mathsf{T} + n^{1+\omega})\mathsf{M}(2^k)\big)$ arithmetic operations in $\mathbb{K}$ (see, e.g., [18], [7]). With a similar cost we compute the evaluation $Adj\big(J_F(\Psi_k)\big)$ of the adjoint matrix of J_F at Ψ_k and the product $\det J_F(\Psi_k)^{-1} \cdot Adj\big(J_F(\Psi_k)\big)$.

Thus, the computation of Ψ_k for every $2 \leq k \leq \kappa$ requires

$$O\Big((\mathsf{T} + n^{1+\omega}) \sum_{k=0}^{\kappa-1} \mathsf{M}(2^k)\Big) = O((\mathsf{T} + n^{1+\omega})\mathsf{M}(D\delta))$$

arithmetic operations in $\mathbb{K}$. The remaining steps do not change the overall asymptotic cost. □

Next we discuss how we can solve the Hermite–Padé approximation problem (5.3). This is represented by a linear system with $N+1$ equations and $O(D\delta)$ unknowns, given by the coefficients of the solution $g \in \mathbb{K}[S,T]$ of (5.2). Best general–purpose algorithms solving a system of size $O(D\delta \times D\delta)$ require $O((D\delta)^\omega)$ arithmetic operations [7]. However, in this case we profit from the structure of (5.3): it turns out that for a suitable ordering of the unknowns, the matrix M of the system (5.3) is a block–Toeplitz matrix (see, e.g., [12, Lemma 4.3], [8, Lemma 6]). This allows us to solve (5.3) using the theory of matrices of fixed displacement rank (cf. [7], [33]). We shall apply the algorithm of [8], which is aimed at solving linear systems defined by matrices of "large" displacement rank.

Further, as we are interested in the least-degree nonzero solution $g \in \mathbb{K}[S,T]$ of (5.3), we combine [8] with a strategy of binary search as in, e.g., [7, Algorithm 8.2]. Fix $\rho \le \delta$. From [8, Corollary 1] it follows that, if there exist nonzero solutions $g \in \mathbb{K}[S,T]$ of (5.3) with $\deg_S g \le \rho$, then one such solution can be computed with $O(D^{\omega-1}\mathsf{M}(D\rho)\log(D\rho))$ arithmetic operations in $\mathbb{K}$ and error probability at most $2(D\rho)^2/|\mathbb{K}|$. Therefore, applying a binary search we can determine the least-degree solution of (5.3) with at most $\lceil \log \delta \rceil$ such steps. As a consequence, we have the following result:

PROPOSITION 5.2. *Suppose that we are given the dense representation of the powers $\sigma_N, \ldots, \sigma_N^D$, as provided by the algorithm underlying Proposition 5.1. Then the minimal polynomial $m_S \in \mathbb{K}[S,T]$ can be computed with $O(D^{\omega-1}\mathsf{M}(D\delta)\log^2\delta)$ operations in $\mathbb{K}$ and error probability at most $2(D\delta)^2\log\delta/|\mathbb{K}|$.*

Combining Propositions 5.1 and 5.2 we obtain an algorithm computing the minimal polynomial m_S from the rational functions $F_1, \ldots, F_n$:

PROPOSITION 5.3. *The polynomial $m_S \in \mathbb{K}[S,T]$ can be computed with $O((\mathsf{T} + n^{1+\omega} + D^{\omega-1}\log^2\delta)\mathsf{M}(D\delta))$ operations in $\mathbb{K}$ and error probability at most $2(D\delta)^2\log\delta/|\mathbb{K}|$.*

5.2. A geometric solution of $\mathcal{C}$. Our next task consists in extending the algorithm of the previous section to an algorithm computing a geometric solution of the curve $\mathcal{C}$ defined in (5.1). Let $\Lambda := (\Lambda_1, \ldots, \Lambda_n)$ be a vector of new indeterminates and consider the projection map $\pi_\Lambda : \mathbb{A}^n \times \mathcal{C} \to \mathbb{A}^n \times \mathbb{A}^1$ defined by $\pi_\Lambda(\lambda, s, x) := (\lambda, s)$. Since π_S is a dominant morphism, so is π_Λ and $\mathbb{K}(\Lambda, S) \hookrightarrow \mathbb{K}(\Lambda) \otimes_{\mathbb{K}} \mathbb{K}(\mathcal{C})$ is an algebraic field extension. The minimal polynomial $m_\Lambda \in \mathbb{K}[\Lambda, S, T]$ of the linear form $\mathcal{L}_\Lambda := \Lambda \cdot X$ in $\mathbb{K}(\Lambda, S) \hookrightarrow \mathbb{K}(\Lambda) \otimes_{\mathbb{K}} \mathbb{K}(\mathcal{C})$ is a separable element of $\mathbb{K}[\Lambda, S][T]$ satisfying the degree bounds $\deg_T m_\Lambda \le D$, $\deg_S m_\Lambda \le \delta$ and $\deg_\Lambda m_\Lambda \le \delta$ (see, e.g., [11, Proposition 6.1] and [39, Proposition 1]). Notice that substituting λ for Λ we have $m_\Lambda(\lambda, S, T) = m_S(S,T)$.

Applying the algorithm underlying Proposition 5.3 to the linear form $\mathcal{L}_\Lambda$ we compute the minimal polynomial $m_\Lambda(\Lambda, S, T)$ with $O((\mathsf{T} + n^{1+\omega} + D^{\omega-1}\log^2\delta)\ \mathsf{M}(D\delta))$ arithmetic operations in $\mathbb{K}(\Lambda)$. Therefore, by Lemma 3.1 we obtain the following result:

PROPOSITION 5.4. *Suppose that the coefficients of the linear form $\mathcal{L}$ are randomly chosen in $\mathbb{K}$. Then we can compute a geometric solution of $\mathcal{C}$ with $O((\mathsf{T} + n^{1+\omega} + D^{\omega-1}\log^2\delta)n\mathsf{M}(D\delta))$ operations in $\mathbb{K}$. Furthermore, the algorithm output the right result with error probability at most $9\mathcal{D}\delta\log\delta/|\mathbb{K}|$, where $\mathcal{D} := D(\delta+2)^{\log\delta}$.*

Proof. As explained in the proof of Lemma 3.1, we apply the algorithm for the computation of the minimal polynomial m_S of Proposition 5.3 to the generic linear form $\mathcal{L}_\Lambda$, truncating each intermediate result up to order $(\Lambda-\lambda)^2$. Therefore, from Lemma 3.1 and Proposition 5.3 we easily deduce the complexity estimate of the proposition.

In order to estimate the error probability of the algorithm we have to estimate the probability of failure of the choice of the vector of coefficients of the linear form $\mathcal{L}$. Recall that the application of Lemma 3.1 requires that the vector of coefficients $\lambda := (\lambda_1, \ldots, \lambda_n)$ of the linear form $\mathcal{L}$ does not annihilate any denominator in $\mathbb{K}[\Lambda]$ of any intermediate result of the algorithm computing the minimal polynomial m_Λ.

The algorithm for obtaining the polynomial m_Λ consists of two steps: the computation of the first D powers of $\sigma_N^{(\Lambda)} := \mathcal{L}_\Lambda(\Psi_k)$, which is considered in Proposition 5.1, and the solution of the Hermite–Padé approximation problem (5.3), which is considered in Proposition 5.2. From Algorithm 5.2 we conclude that the computation of the powers $\sigma_N^{(\Lambda)}, \ldots, (\sigma_N^{(\Lambda)})^D$ does not require any division by a nonconstant polynomial of $\mathbb{K}[\Lambda]$.

Next, we analyze the divisions necessary to solve the Hermite–Padé approximation problem (5.3), which is solved applying an algorithm of [8]. This algorithm is an adaptation of Kaltofen's *Leading Principal Inverse* algorithm ([26], [27]). Kaltofen's algorithm performs a recursive reduction of the computation of the inverse of a "generic–rank–profile" square input matrix

$$A = \begin{pmatrix} A_{1,1} & A_{1,2} \\ A_{2,1} & A_{2,2} \end{pmatrix}$$

to that of the leading principal submatrix $A_{1,1}$ and its Schur complement $\Delta := A_{2,2} - A_{2,1}A_{1,1}^{-1}A_{1,2}$. The divisions which arise during the execution of this recursive step are related to the computation of $A_{1,1}^{-1}$ and Δ^{-1} and a routine of "compression" (cf. [7, Problem 2.2.11.c]) of the generators of matrices which are obtained as certain products involving $A_{1,1}^{-1}$, Δ^{-1}, $A_{1,2}$ and $A_{2,1}$. The latter in turn requires the computation of the inverses of certain submatrices of the products under consideration.

Each entry of the matrix M of the linear system (5.3) is a coefficient of a power $(\sigma_N^{(\Lambda)})^j$, which is therefore a polynomial in Λ of degree at most $j \leq D$. Since the generic–rank–profile matrix A is obtained by multiplying M with suitable matrices with entries in $\mathbb{K}$, we conclude that the entries of $A_{1,1}$ are polynomials of $\mathbb{K}[\Lambda]$ of degree at most D, while the numerators and denominators of the entries of Δ^{-1} are polynomials of $\mathbb{K}[\Lambda]$ of degree at most $D(\delta+2)$. Therefore, by a simple recursive argument we see that the numerators and denominators of all leading principal submatrices and Schur complements which are inverted during the algorithm have degrees bounded by $\mathcal{D} := D(\delta+2)^{\log\delta}$. This in turn implies that the denominators arising during the compression routine have degrees bounded by $3\mathcal{D}\delta$.

Finally, taking into account that the algorithm of [8] consists of at most $\lfloor \log \delta \rfloor$ recursive steps, and that each recursive step requires the inversion of at most 4 matrices, we conclude that the product of all the denominators arising during the algorithm has degree bounded by $8\mathcal{D}\delta\log\delta$. By Corollary 2.1, it follows that a random choice of $\lambda := (\lambda_1, \ldots, \lambda_n)$ does not vanish any denominator of the algorithm computing the minimal polynomial m_Λ with error probability at most $8\mathcal{D}\delta\log\delta/|\mathbb{K}|$. Putting together this estimate and the error probability $2(D\delta)^2\log\delta/|\mathbb{K}|$ of the algorithm underlying Proposition 5.3 we deduce the statement of the proposition. □

5.3. Computation of the points of $F^{-1}(y^{(0)}) \cap \mathbb{F}_q^n$. In this section we show how to find the solutions in $\mathbb{F}_q^n$ of the system $F(X) = y^{(0)}$.

Assume that we are given a geometric solution defined over $\mathbb{K}$ of the curve C defined in (5.1), as provided by the algorithm of Proposition 5.4. This geometric solution consists of a linear form $\mathcal{L} \in \mathbb{K}[X]$, the minimal polynomial $m_S \in \mathbb{K}[S,T]$ of $\mathcal{L}$ in the algebraic field extension $\mathbb{K}(S) \hookrightarrow \mathbb{K}(C)$ and the parametrizations $(\partial m_S/\partial T)X_j - v_j(S,T)$ $(1 \le j \le n)$ of the variables $X_1, \ldots, X_n$ by the zeros of m_S.

Let $\pi_S^{-1}(0) = \{0\} \times C_0$, where $C_0 := F^{-1}(y^{(0)})$. Since $\mathcal{L}$ separates the points of $\pi_S^{-1}(0)$, from a geometric solution of C we obtain a geometric solution of C_0. Indeed, substituting 0 for S in $m_S, v_1, \ldots, v_n$ we obtain polynomials $m_S(0,T), v_1(0,T), \ldots, v_n(0,T) \in \mathbb{K}[T]$ which represent a complete description of the fiber C_0, i.e., we have the identity

$$C_0 = \{x \in \mathbb{A}^n; m_S(0,\mathcal{L}(x)) = 0, m_S'(0,\mathcal{L}(x))x_j = v_j(0,\mathcal{L}(x))\ (1 \le j \le n)\},$$

where $m_S'(0,T) := \frac{\partial m_S}{\partial T}(0,T)$. Nevertheless, the polynomials $m_S(0,T)$, $v_1(0,T), \ldots, v_n(0,T) \in \mathbb{K}[T]$ do not necessarily form a geometric solution of C_0, because the polynomial $m_S(0,T)$ might have multiple factors. In such a case, it is easy to see that the multiple factors of $m_S(0,T)$ are also factors of $v_1(0,T), \ldots, v_n(0,T)$ (see, e.g., [21, §6.5]). To remove these multiple factors, we proceed in the following way: first, we compute

$$a(T) := \gcd\big(m_S(0,T), m_S'(0,T)\big), \quad m_0(T) := \frac{m_S(0,T)}{a(T)},$$

which yield the square–free representation $m_0(T)$ of $m_S(0,T)$. Next, given that $a(T)$ divides $v_j(0,T)$ for $1 \le j \le n$, we obtain polynomials

$$\frac{m_S'(0,T)}{a(T)}X_j - \frac{v_j(0,T)}{a(T)} \quad (1 \le j \le n),$$

which vanish on the points of C_0. Finally, since m_0 and $m_S'(0,T)/a(T)$ have no common factors in $\mathbb{K}[T]$, we invert $m_S'(0,T)/a(T)$ modulo $m_0(T)$ and obtain parametrizations $X_j - w_j(T)$ $(1 \le j \le n)$ of the coordinates of the points of C_0 by the zeros of $m_0(T)$ which are better suited for our purposes. In the next lemma we state the cost of this procedure:

LEMMA 5.2. *Given a geometric solution of the curve C, as provided by the algorithm underlying Proposition 5.4, we can deterministically compute a geometric solution $m_0(T), X_1 - w_1(T), \ldots, X_n - w_n(T)$ of the zero dimensional variety C_0 with $O(n\delta\mathsf{M}(D))$ operations in $\mathbb{K}$.*

Proof. The dense representation of the polynomials $m_S(0,T), v_1(0,T), \ldots, v_n(0,T)$ can be obtained from the dense representation of the polynomials $m_S(S,T), v_1(S,T), \ldots, v_n(S,T)$ with $O(nD\delta)$ arithmetic operations in $\mathbb{K}$. The remaining computations are $O(n)$ multiplications, greatest common divisors and a modular inversion of univariate polynomials, whose degrees are less than or equal to D, which contribute with $O\big(n\mathsf{M}(D)\big)$ additional arithmetic operations in $\mathbb{K}$. □

Finally, we compute the $\mathbb{K}$–rational points of C_0, which in particular yield the solutions in $\mathbb{F}_q^n$ of $F(X) = y^{(0)}$. For this purpose, set

$$h := \gcd(m_0, T^{|\mathbb{K}|} - T) \in \mathbb{K}[T].$$

Following, e.g., [18, Corollary 14.16], we have that h can be computed with $O\big(\mathsf{M}(D)\log|\mathbb{K}|\big)$ arithmetic operations in $\mathbb{K}$. Since h factors into linear factors, its factorization can be computed with $O\big(\mathsf{M}(D)\log(\mu D|\mathbb{K}|)\big)$ arithmetic operations in $\mathbb{K}$ and error probability at most $1/\mu$ (see, e.g., [18, Theorem 14.9]).

Observe that the roots of h are the values $\mathcal{L}(P)$ resulting from the evaluation of $\mathcal{L}$ in the points $P \in C_0 \cap \mathbb{K}^n$. In particular, $\mathcal{L}(x) \in \mathbb{K}$ is a root of h for every point $x \in C_0 \cap \mathbb{K}^n$. Thus, if we substitute the roots α of h for T in the polynomials $w_j(T)$ $(1 \le j \le n)$, we obtain all the points of $C_0 \cap \mathbb{F}_q^n$ as the points $\big(w_1(\alpha), \ldots, w_n(\alpha)\big) \in \mathbb{F}_q^n$ with $h(\alpha) = 0$. Since such substitutions require $O(nD)$ additional arithmetic operations in $\mathbb{K}$, we have the following result:

PROPOSITION 5.5. *Given a geometric solution of the zero-dimensional variety $C_0 = F^{-1}(y^{(0)})$, as provided by the algorithm underlying Lemma 5.2, we can compute the set $C_0 \cap \mathbb{F}_q^n$ with $O\big(\mathsf{M}(D)(n + \log(\mu D|\mathbb{K}|))\big)$ arithmetic operations in $\mathbb{K}$ and error probability at most $1/\mu$.*

Putting together Propositions 5.4 and 5.5 we obtain our main result:

THEOREM 5.3. *The solutions in $\mathbb{F}_q^n$ of the input system $F(X) = y^{(0)}$ can be computed with*

$$O\Big(\big((\mathsf{T}+n^{1+\omega}+D^{\omega-1}\log^2\delta)\mathsf{M}(D\delta)+\mathsf{M}(D)\mathsf{M}(\log q+\log^2\delta)\big)n\mathsf{M}(\log q+\log^2\delta)\Big)$$

bit operations and error probability at most $1/4$.

Proof. Let $\mathbb{K}$ be a field extension of $\mathbb{F}_q$ of cardinality greater than $128\mathcal{D}\delta\log\delta$, where $\mathcal{D} := D(\delta+2)^{\log\delta}$. Choose randomly a point $(\lambda, x^{(1)}) \in \mathbb{K}^{2n}$ and set $\mathcal{L} := \lambda \cdot X \in \mathbb{K}[X]$. Suppose that $(\lambda, x^{(1)}) \in \mathbb{K}^{2n}$ satisfies the following conditions:

1. $y^{(1)} := F(x^{(1)})$ is a lifting point of $\pi : V \to \mathbb{A}^n$,
2. the curve $C \subset \mathbb{A}^{n+1}$ defined in (5.1) is absolutely irreducible,
3. the linear form $\mathcal{L}$ separates the fibers $V_{y^{(1)}}$ and $V_{y^{(0)}}$,
4. λ does not annihilate any denominator of the algorithm computing the minimal polynomial m_Λ underlying Proposition 5.3.

Applying the algorithm of Proposition 5.4 we obtain a geometric solution of the curve C with $O\big((\mathsf{T} + n^{1+\omega} + D^{\omega-1}\log^2\delta)n\mathsf{M}(D\delta)\big)$ operations in $\mathbb{K}$.

Then we apply the algorithm of Proposition 5.5 in order to compute the solutions in $\mathbb{F}_q^n$ of $F(X) = y^{(0)}$ with $O(\mathsf{M}(D)(n + \log(D|\mathbb{K}|)))$ arithmetic operations in $\mathbb{K}$. Combining both complexity estimates, and taking into account that every arithmetic operation in $\mathbb{K}$ requires $O(\mathsf{M}(\log^2 \delta + \log q))$ bit operations, we deduce the estimate of the statement of the theorem.

By Corollary 4.1 we have that $(\lambda, x^{(1)})$ satisfies conditions (1)–(3) above with error probability at most $4d\delta^4/|\mathbb{K}| \leq 1/16$. Furthermore, from the proof of Proposition 5.3 we conclude that condition (4) is satisfied with error probability at most 1/16. Finally, assuming that (1)–(4) hold, the algorithms underlying Propositions 5.3 and 5.5 output the right results with error probability at most $2(D\delta)^2 \log \delta/|\mathbb{K}| \leq 1/16$ and 1/16 (fixing $\mu := 16$ in Proposition 5.5) respectively. This shows that the overall error probability is at most 1/4 and finishes the proof of the theorem. □

We make a few remarks concerning Theorem 5.3. Observe that our algorithm has a cost of $O(n^{2+\omega}D^\omega \delta \log q + nD \log^2 q)$ bit operations, up to logarithmic terms. This improves and extends the algorithm of [12]. We have further contributed to the latter by providing estimates for the corresponding error probability.

A second remark concerns the behavior of our algorithm under the hypotheses of [41]. Recall that [41] requires $F : \mathbb{A}^n \to \mathbb{A}^n$ to be a polynomial map which is polynomially invertible, with inverse $G := (G_1, \ldots, G_n)$ with degrees polynomially bounded with respect to n and $d := \max_{1 \leq k \leq n} \deg F_k$. Under these hypotheses, the authors exhibit an algorithm which computes the inverse mapping G with a polynomial cost in T, n and d. Under these conditions, we have that the projection mapping $\pi : V \to \mathbb{A}^n$ has degree 1, i.e., the identity $D = 1$ holds. Furthermore, it is easy to see that the minimal polynomial $m_S(S, T)$ has degree bounded by $e := \max_{1 \leq k \leq n} \deg G_k$. Therefore, the algorithms underlying Propositions 5.4 and 5.5 have actually polynomial cost in T, n and e. This shows that the cost of our algorithm meets this polynomial bound assuming the strong hypotheses of [41].

6. Conclusions. Our complexity estimate may be roughly described as polynomial in the cost T of the evaluation of the input rational functions $F_1, \ldots, F_n$, the number of variables n, the logarithm $\log q$ of the cardinality of the field $\mathbb{F}_q$ and two geometric invariants: the degree D of the mapping F and the degree δ of the graph of F. In this sense, we see that the practical convenience of our algorithm, and the subsequent (in)security of cryptosystems based on polynomial or rational mappings over a finite field, essentially relies on these geometric invariants.

In worst case we have $D = \delta = \deg(F_1) \cdots \deg(F_n)$, which implies that our algorithm is exponential in the input size. Furthermore, adapting the arguments of [13] it is possible to prove that any *universal* algorithm solving $F(X) = y^{(0)}$ has necessarily cost $(\deg(F_1) \cdots \deg(F_n))^{\Omega(1)}$, showing thus the security of the corresponding cryptosystem with respect to *universal* decoding algorithms. Since a universal algorithm is one which does not distinguish input systems according to geometric invariants and represents a model for the standard algorithms based on rewriting techniques, such as Gröbner basis algorithms, such cryptosystems are likely to be secure.

REFERENCES

[1] M. Alonso, E. Becker, M.-F. Roy, and T. Wörmann, *Zeroes, multiplicities and idempotents for zerodimensional systems*, in Proceedings of MEGA'94, Vol. **143** of Progr. Math., Boston, 1996, Birkhäuser, pp. 1–15.

[2] J. Balcázar, J. Díaz, and J. Gabarró, *Structural complexity I*, Vol. **11** of Monogr. Theoret. Comput. Sci. EATCS Ser., Springer, Berlin, 1988.

[3] M. Bardet, *Etude des systèmes algébriques surdétermines. Applications aux codes correcteurs et à la cryptographie*, PhD thesis, Université Paris **6**, 2004.

[4] M. Bardet, J.-C. Faugère, and B. Salvy, *Complexity of Gröbner basis computation for semi-regular overdetermined sequences over $\mathbb{F}_2$ with solutions in $\mathbb{F}_2$*. Rapport de Recherche INRIA RR-5049, `www.inria.fr/rrrt/rr-5049.html`, 2003.

[5] W. Baur and V. Strassen, *The complexity of partial derivatives*, Theoret. Comput. Sci., **22** (1983), pp. 317–330.

[6] E. Biham and A. Shamir, *Differential cryptanalysis of DES-like cryptosystems*, J. Cryptology, **4** (1991), pp. 3–72.

[7] D. Bini and V. Pan, *Polynomial and matrix computations*, Progress in Theoretical Computer Science, Birkhäuser, Boston, 1994.

[8] A. Bostan, C.-P. Jeannerod, and E. Schost, *Solving Toeplitz- and Vandermonde-like linear systems with large displacement rank.* To appear in Proceedings ISSAC'07, `http://www-sop.inria.fr/saga/POL`, 2007.

[9] P. Bürgisser, M. Clausen, and M. Shokrollahi, *Algebraic Complexity Theory*, Vol. **315** of Grundlehren Math. Wiss., Springer, Berlin, 1997.

[10] A. Cafure and G. Matera, *Fast computation of a rational point of a variety over a finite field*, Math. Comp., **75** (2006), pp. 2049–2085.

[11] ———, *Improved explicit estimates on the number of solutions of equations over a finite field*, Finite Fields Appl., **12** (2006), pp. 155–185.

[12] A. Cafure, G. Matera, and A. Waissbein, *Inverting bijective polynomial maps over finite fields*, in Proceedings of the 2006 Information Theory Workshop, ITW2006, G. Seroussi and A. Viola, eds., IEEE Information Theory Society, 2006, pp. 27–31.

[13] D. Castro, M. Giusti, J. Heintz, G. Matera, and L.M. Pardo, *The hardness of polynomial equation solving*, Found. Comput. Math., **3** (2003), pp. 347–420.

[14] N. Courtois, A. Klimov, J. Patarin, and A. Shamir, *Efficient algorithms for solving overdefined systems of multivariate polynomial equations*, in EUROCRYPT 2000, B. Preneel, ed., Vol. 1807 of Lecture Notes in Comput. Sci., Berlin, 2000, Springer, pp. 71–79.

[15] C. de Cannière, A. Biryukov, and B. Preneel, *An introduction to block cipher cryptanalysis*, Proc. IEEE, **94** (2006), pp. 346–356.

[16] J.-C. Faugère, *A new efficient algorithm for computing Gröbner bases without reduction to zero (F5)*, Proceedings ISSAC'02, T. Mora, ed., New York, 2002, ACM Press, pp. 75–83.

[17] S. Gao, *Factoring multivariate polynomials via partial differential equations*, Math. Comp., **72** (2003), pp. 801–822.

[18] J. von zur Gathen and J. Gerhard, *Modern computer algebra*, Cambridge Univ. Press, Cambridge, 1999.

[19] M. Garey and D. Johnson, *Computers and Intractability: A Guide to the Theory of NP-Completeness*, Freeman, San Francisco, 1979.

[20] M. Giusti, K. Hägele, J. Heintz, J.E. Morais, J.L. Montaña, and L.M. Pardo, *Lower bounds for Diophantine approximation*, J. Pure Appl. Algebra, **117, 118** (1997), pp. 277–317.

[21] M. GIUSTI, G. LECERF, AND B. SALVY, *A Gröbner free alternative for polynomial system solving*, J. Complexity, 17 (2001), pp. 154–211.

[22] J. HEINTZ, *Definability and fast quantifier elimination in algebraically closed fields*, Theoret. Comput. Sci., **24** (1983), pp. 239–277.

[23] M.-D. HUANG AND Y.-C. WONG, *Solvability of systems of polynomial congruences modulo a large prime*, Comput. Complexity, **8** (1999), pp. 227–257.

[24] H. IMAI AND T. MATSUMOTO, *Public quadratic polynomial-tuples for efficient signature-verification and message-encryption*, in Advances in Cryptology - EUROCRYPT '88, C. Günther, ed., Vol. **330** of Lecture Notes in Comput. Sci., Berlin, 1988, Springer, pp. 419–453.

[25] J.-R. JOLY, *Equations et variétés algébriques sur un corps fini*, Enseign. Math., **19** (1973), pp. 1–117.

[26] E. KALTOFEN, *Asymptotically fast solution of Toeplitz–like singular linear systems*, in Proceedings ISSAC'94, J. von zur Gathen and M. Giesbrecht, eds., New York, 1994, ACM Press, pp. 297–304.

[27] ———, *Analysis of Coppersmith's block Wiedemann algorithm for the parallel solution of sparse linear systems*, Math. Comp., **64** (1995), pp. 777–806.

[28] ———, *Effective Noether irreducibility forms and applications*, J. Comput. System Sci., **50** (1995), pp. 274–295.

[29] A. KIPNIS AND A. SHAMIR, *Cryptanalysis of the HFE Public Key Cryptosystem by relinearization*, in Advances in Cryptology – CRYPTO'99, M. Wiener, ed., Vol. **1666** of Lecture Notes in Comput. Sci., Berlin, 1999, Springer, pp. 19–30.

[30] N. KOBLITZ, *Algebraic aspects of cryptography*, Vol. **3** of Algorithms Comput. Math., Springer, Berlin Heidelberg New York, corrected 2nd printing ed., 1999.

[31] G. LECERF, *Improved dense multivariate polynomial factorization algorithms*, J. Symbolic Comput., **42** (2007), pp. 477–494.

[32] R. LIDL AND H. NIEDERREITER, *Finite fields*, Addison–Wesley, Reading, Massachusetts, 1983.

[33] V. PAN, *Structured matrices and polynomials. Unified superfast algorithms*, Birkhäuser, Boston, 2001.

[34] L.M. PARDO AND J. SAN MARTÍN, *Deformation techniques to solve generalized Pham systems*, Theoret. Comput. Sci., **315** (2004), pp. 593–625.

[35] J. PATARIN, *Cryptoanalysis of the Matsumoto and Imai Public Key Scheme of Eurocrypt'88*, in Advances in Cryptology - CRYPTO '95, D. Coppersmith, ed., Vol. **963** of Lecture Notes in Comput. Sci., Springer, 1995, pp. 248–261.

[36] ———, *Asymmetric cryptography with a hidden monomial*, in Advances in Cryptology - CRYPTO '96, N. Koblitz, ed., Vol. **1109** of Lecture Notes in Comput. Sci., Springer, 1996, pp. 45–60.

[37] F. ROUILLIER, *Solving zero–dimensional systems through rational univariate representation*, Appl. Algebra Engrg. Comm. Comput., **9** (1997), pp. 433–461.

[38] J. SAVAGE, *Models of Computation. Exploring the Power of Computing*, Addison Wesley, Reading, Massachussets, 1998.

[39] E. SCHOST, *Computing parametric geometric resolutions*, Appl. Algebra Engrg. Comm. Comput., **13** (2003), pp. 349–393.

[40] I. SHAFAREVICH, *Basic Algebraic Geometry: Varieties in Projective Space*, Springer, Berlin Heidelberg New York, 1994.

[41] C. STURTIVANT AND Z.-L. ZHANG, *Efficiently inverting bijections given by straight line programs*, in Proceedings of the 31st Annual Symp. Found. Comput. Science, FOCS'90, Vol. **1**, IEEE Computer Society Press, 1990, pp. 327–334.

[42] L.-C. WANG AND F.-H. CHANG, *Tractable rational map cryptosystem*. Cryptology ePrint Archive, Report 2004/046, `http://eprint.iacr.org/2004/046/`, 2004.

[43] C. WOLF AND B. PRENEEL, *Taxonomy of public key schemes based on the problem of multivariate quadratic equations*. Cryptology ePrint Archive, Report 2005/077, `http://eprint.iacr.org/2005/077/`, 2005.

HIGHER-ORDER DEFLATION FOR POLYNOMIAL SYSTEMS WITH ISOLATED SINGULAR SOLUTIONS

ANTON LEYKIN*, JAN VERSCHELDE†, AND AILING ZHAO†‡

Abstract. Given an approximation to a multiple isolated solution of a system of polynomial equations, we provided a symbolic-numeric deflation algorithm to restore the quadratic convergence of Newton's method. Using first-order derivatives of the polynomials in the system, our first-order deflation method creates an augmented system that has the multiple isolated solution of the original system as a regular solution.

In this paper we consider two approaches to computing the "multiplicity structure" at a singular isolated solution. An idea coming from one of them gives rise to our new higher-order deflation method. Using higher-order partial derivatives of the original polynomials, the new algorithm reduces the multiplicity faster than our first method for systems which require several first-order deflation steps. In particular: the number of higher-order deflation steps is bounded by the number of variables.

Key words. Deflation, isolated singular solutions, Newton's method, multiplicity, polynomial systems, reconditioning, symbolic-numeric computations.

AMS(MOS) subject classifications. Primary 65H10. Secondary 14Q99, 68W30.

1. Introduction.

This paper describes a numerical treatment of singular solutions of polynomial systems. A trivial example to consider would be a single equation with a double root, $f(x) = x^2 = 0$, or a cluster of two very close roots, $f(x) = x^2 - \varepsilon^2 = 0$, where $0 < \varepsilon \ll$ machine precision. In both cases getting good approximate solutions with straightforward numerical approaches such as Newton's method is not easy. Instead of attempting to solve the given equations we replace them with the system augmented by the equation's derivative, $\bar{f}(x) = (f(x), f'(x)) = \mathbf{0}$. Note that this completely symbolic procedure leads to a system with exact regular roots in the first case, whereas in the second case the system $\bar{f}(x) = \mathbf{0}$ is inconsistent. However, a numerical solver applied to the latter converges to a regular solution of a close-by system.

In general setting, given a square or overdetermined system of equations in many variables with a multiple isolated solution (a cluster of solutions) our approach deflates the multiplicity of the solution (cluster) by applying a certain numerical procedure. From the point of view of

*Institute for Mathematics and its Applications, University of Minnesota, 207 Church Street S.E., Minneapolis, MN 55455-0436, USA. (leykin@ima.umn.edu; http://www.ima.umn.edu/~leykin).

†Department of Mathematics, Statistics, and Computer Science, University of Illinois at Chicago, 851 South Morgan (M/C 249), Chicago, IL 60607-7045, USA (jan@math.uic.edu or jan.verschelde@na-net.ornl.gov; http://www.math.uic.edu/~jan). This material is based upon work supported by the National Science Foundation under Grant No. 0134611 and Grant No. 0410036.

‡azhao1@uic.edu; http://www.math.uic.edu/~azhao1.

numerical analysis it may be called a *reconditioning* method: to recondition means to reformulate a problem so its condition number improves.

Our deflation method was first presented at [29], and then described in greater detail in [16]. In [15], a directed acyclic graph of Jacobian matrices was introduced for an efficient implementation. We will call the deflation of [16] the first-order deflation to distinguish it from the higher-order deflation proposed in this paper.

On input we consider clusters of approximate zeroes of systems $F(\mathbf{x}) = (f_1(\mathbf{x}), f_2(\mathbf{x}), \ldots, f_N(\mathbf{x})) = \mathbf{0}$ of N equations in n unknowns $\mathbf{x} \in \mathbb{C}^n$. We assume the cluster approximates an isolated solution $\mathbf{x}^*$ of $F(\mathbf{x}) = \mathbf{0}$. Therefore, $N \geq n$. As $\mathbf{x}^*$ is a singular solution, the Jacobian matrix of $F(\mathbf{x})$, denoted by $A(\mathbf{x})$, is singular at $\mathbf{x}^*$. In particular, we have $r = \text{Rank}(A(\mathbf{x}^*)) < n$.

In case $r = n - 1$, consider a nonzero vector $\boldsymbol{\lambda}$ in the kernel of $A(\mathbf{x}^*)$, which we denote by $\boldsymbol{\lambda} \in \ker(A(\mathbf{x}^*))$, then the equations

$$g_i(\mathbf{x}) = \sum_{j=1}^{n} \lambda_j \frac{\partial f_i(\mathbf{x})}{\partial x_j}, \quad i = 1, 2, \ldots, N, \tag{1.1}$$

vanish at $\mathbf{x}^*$, because $r = \text{Rank}(A(\mathbf{x}^*)) < n$. For $r < n-1$, our algorithm reduces to the corank-1 case, replacing $A(\mathbf{x})$ by $A(\mathbf{x})B$, where B is a random complex N-by-$(r+1)$ matrix. For the uniqueness $\boldsymbol{\lambda} \in \ker(A(\mathbf{x}^*))$, we add a linear scaling equation $\langle \mathbf{h}, \boldsymbol{\lambda} \rangle = 1$ (using a random complex $(r+1)$-vector $\mathbf{h}$). and consider the augmented system

$$G(\mathbf{x}, \boldsymbol{\lambda}) = \begin{cases} F(\mathbf{x}) & = \mathbf{0} \\ A(\mathbf{x})B\boldsymbol{\lambda} & = \mathbf{0} \\ \langle \mathbf{h}, \boldsymbol{\lambda} \rangle & = 1. \end{cases} \tag{1.2}$$

Let us denote by $\mu_F(\mathbf{x}^*)$ the multiplicity of $\mathbf{x}^*$ as a solution of the system $F(\mathbf{x}) = 0$. In [16] we proved that there is a $\boldsymbol{\lambda}^*$ such that $\mu_G(\mathbf{x}^*, \boldsymbol{\lambda}^*) < \mu_F(\mathbf{x}^*)$. Therefore, our first-order deflation algorithm takes at most $m-1$ stages to determine $\mathbf{x}^*$ as a regular root of an augmented polynomial system.

Related work. The literature on Newton's method is vast. As stated in [5], Lyapunov-Schmidt reduction (see also [1], [8, §6.2], [13], [17], and [19, §6.1]) stands at the beginning of every mathematical treatment of singularities. We found the inspiration to develop a symbolic-numeric deflation algorithm in [22]. The symbolic deflation procedure of [14] restores the quadratic convergence of Newton's method with a complexity proportional to the square of the multiplicity of the root. Smale's α-theory is applied to clusters of zeroes of analytic functions in [6] and to special multivariate cases in [7]. Algorithms to compute the multiplicity are presented in [2], [3], and [26].

Outline of this paper. We sketch the link between two different objects describing what we call the *multiplicity structure* of an isolated singular solution: the dual space of differential functionals and the initial ideal with respect to a local monomial order, both associated to the ideal generated by the polynomials in the system in the polynomial ring.

Next, following the latter method, we explain how to compute a basis of the dual space, first, following the ideas of Dayton and Zeng [3], then using the approach of Stetter and Thallinger [26]. We provide a formal symbolic algorithm for each approach, respectively called the DZ and ST algorithms; the ingredients of the algorithms do not go beyond linear algebra. Moreover, we present an algorithm to determine the order of the deflation.

The formalism developed for DZ and ST algorithms found a natural continuation in the *higher-order deflation* method that generalizes and extends the first-order deflation in [16]. For the systems that require more than one deflation step by our first algorithm, the new deflation algorithm is capable of completing the deflation in fewer steps.

2. Statement of the main theorem & algorithms. The matrices $A^{(d)}(x)$ we introduce below coincide for $d = 1$ with the Jacobian matrix of a polynomial system. They are generalizations of the Jacobian matrix, built along the same construction as the matrices used in the computation of the multiplicity by Dayton and Zeng in [3].

DEFINITION 2.1. The *deflation matrix* $A^{(d)}(x)$ *of a polynomial system* $F = (f_1, f_2, \ldots, f_N)$ of N equations in n unknowns $x = (x_1, x_2, \ldots, x_n)$ is a matrix with elements in $\mathbb{C}[x]$. The rows of $A^{(d)}(x)$ are indexed by $x^\alpha f_j$, where $|\alpha| < d$ and $j = 1, 2, \ldots, N$. The columns are indexed by partial differential operators $\partial^\beta = \frac{\partial^{|\beta|}}{\partial x_1^{\beta_1} \cdots \partial x_n^{\beta_n}}$, where $\beta \neq \mathbf{0}$ and $|\beta| \leq d$. The element at row $x^\alpha f_j$ and column ∂^β of $A^{(d)}(x)$ is

$$\partial^\beta \cdot (x^\alpha f_j) = \frac{\partial^{|\beta|}(x^\alpha f_j)}{\partial x^\beta}. \tag{2.1}$$

$A^{(d)}(x)$ has N_r rows and N_c columns, $N_r = N \cdot \binom{n+d-1}{n}$ and $N_c = \binom{n+d}{n} - 1$.

The number d will be referred to as the *order of deflation*

EXAMPLE 1 (Second-order deflation matrix). Consider a system of 3 equations in 2 variables $F = (f_1, f_2, f_3) = \mathbf{0}$, where $f_1 = x_1^2$, $f_2 = x_1^2 - x_2^3$, and $f_3 = x_2^4$. Then the second-order deflation matrix $A^{(2)}(x)$ of F is

$$\begin{array}{c} \\ f_1 \\ f_2 \\ f_3 \\ x_1 f_1 \\ x_1 f_2 \\ x_1 f_3 \\ x_2 f_1 \\ x_2 f_2 \\ x_2 f_3 \end{array} \begin{array}{c} \begin{array}{ccccc} \partial_{x_1} & \partial_{x_2} & \partial_{x_1}^2 & \partial_{x_1}\partial_{x_2} & \partial_{x_2}^2 \end{array} \\ \left[\begin{array}{ccccc} 2x_1 & 0 & 2 & 0 & 0 \\ 2x_1 & -3x_2^2 & 2 & 0 & -6x_2 \\ 0 & 4x_2^3 & 0 & 0 & 12x_2^2 \\ 3x_1^2 & 0 & 6x_1 & 0 & 0 \\ 3x_1^2 & -3x_1x_2^2 & 6x_1 & 3x_2^2 & -6x_1x_2 \\ x_2^4 & 4x_1x_2^3 & 0 & 0 & 0 \\ 2x_1x_2 & x_1^2 & 2x_2 & 2x_1 & 0 \\ 2x_1x_2 & -4x_2^3 & 2x_2 & 2x_1 & -12x_2^2 \\ 0 & 5x_2^4 & 0 & 0 & 20x_2^3 \end{array} \right] \end{array} . \qquad (2.2)$$

Notice that $A^{(1)}(\boldsymbol{x})$ (or the Jacobian matrix of F) is contained in the first three rows and two columns of $A^{(2)}(\boldsymbol{x})$.

DEFINITION 2.2. Let $\boldsymbol{x}^*$ be an isolated singular solution of the system $F(\boldsymbol{x}) = \mathbf{0}$ and let d be the order of the deflation. Take a nonzero N_r-vector $(\lambda_\beta)_{\beta\neq 0,\ |\beta|\leq d}$ in the kernel of $A^{(d)}(\boldsymbol{x}^*)$. It corresponds to what we call a *deflation operator* – a linear differential operator with constant coefficients λ_β

$$Q = \sum_{\beta\neq 0,\ |\beta|\leq d} \lambda_\beta \partial^\beta \in \mathbb{C}[\partial]. \qquad (2.3)$$

We use Q to define N_r new equations

$$g_{j,\alpha}(\boldsymbol{x}) = Q \cdot (\boldsymbol{x}^\alpha f_j) = 0, \quad j = 1, 2, \ldots, N,\ |\alpha| < d. \qquad (2.4)$$

When we consider λ_β as indeterminate, we write $g_{j,\alpha}(\boldsymbol{x})$ as $g_{j,\alpha}(\boldsymbol{x}, \boldsymbol{\lambda})$. In that case, for $m = \operatorname{corank}(A^{(d)}(\boldsymbol{x}^*))$, we define m additional linear equations:

$$h_k(\boldsymbol{\lambda}) = \sum_\beta b_{k,\beta}\lambda_\beta - 1 = 0, \quad k = 1, 2, \ldots, m, \qquad (2.5)$$

where the coefficients $b_{k,\beta}$ are randomly chosen complex numbers. Then we define

$$G^{(d)}(\boldsymbol{x}, \boldsymbol{\lambda}) = \left\{ \begin{array}{rcl} f_j(\boldsymbol{x}) & = & 0, \quad j = 1, 2, \ldots, N; \\ g_{j,\alpha}(\boldsymbol{x}, \boldsymbol{\lambda}) & = & 0, \quad j = 1, 2, \ldots, N,\ |\alpha| < d; \\ h_k(\boldsymbol{\lambda}) & = & 0, \quad k = 1, 2, \ldots, m. \end{array} \right. \qquad (2.6)$$

With (2.6) we end Definition 2.2.

Now we are ready to state our main theorem.

THEOREM 2.1. *Let $\boldsymbol{x}^* \in \mathbb{C}^n$ be an isolated solution of $F(\boldsymbol{x}) = \mathbf{0}$. Consider the system $G^{(d)}(\boldsymbol{x}, \boldsymbol{\lambda}) = \mathbf{0}$ as in (2.6). For a generic choice of coefficients $b_{k,\beta}$, there exists a unique $\boldsymbol{\lambda}^* \in \mathbb{C}^{N_c}$ such that the system*

$G^{(d)}(\boldsymbol{x}, \boldsymbol{\lambda})$ has an isolated solution at $(\boldsymbol{x}^*, \boldsymbol{\lambda}^*)$. Moreover, the multiplicity of $(\boldsymbol{x}^*, \boldsymbol{\lambda}^*)$ in $G(\boldsymbol{x}, \boldsymbol{\lambda}) = \boldsymbol{0}$ is strictly less than the multiplicity of $\boldsymbol{x}^*$ in $F(\boldsymbol{x}) = \boldsymbol{0}$.

REMARK 2.1. Assuming coefficients $(b_{k,\beta})$ are chosen from a complex parameter space, the exceptional set of $(b_{k,\beta})$ that do not produce the conclusion of the Theorem 2.1 is contained in a (closed proper) algebraic subset of this parameter space. By genericity we mean staying away from this exceptional set, which is accomplished by choosing random numbers.

To determine the order d, we propose Algorithm 2.2. This d is then used in Algorithm 2.3.

ALGORITHM 2.2. d = **MinOrderForCorankDrop**$(F, \boldsymbol{x}_0, \varepsilon)$

Input: F is a finite set of polynomials;
$\boldsymbol{x}^0 \approx \boldsymbol{x}^*$, $\boldsymbol{x}^* \in F^{-1}(\boldsymbol{0})$, an isolated multiple solution;
$\varepsilon \geq 0$, a threshold parameter.

Output: d is the minimal number such that the system $G^{(d)}$ given via a generic deflation operator Q of order d has corank of the Jacobian at $\boldsymbol{x}^*$ lower than corank $A(\boldsymbol{x}^*)$.

take a generic vector $\gamma = (\gamma_1, \ldots, \gamma_n) \in \ker A(\boldsymbol{x}^0)$;
let $H(t) = F(\boldsymbol{x}^0 + \gamma t) = F(x_1^0 + \gamma_1 t, \ldots, x_n^0 + \gamma_n t)$;
$d := \min\{\ a \mid \varepsilon <$ coefficient of t^a in $H(t)\ \} - 1$.

See the proof of correctness of this algorithm in the exact setting, i.e., $\boldsymbol{x}^* = \boldsymbol{x}^0$ and $\varepsilon = 0$, in the end of subsection 5.2.

ALGORITHM 2.3. $D^{(d)}F$ = **Deflate**$(F, d, \boldsymbol{x}_0)$

Input: F is a finite set of polynomials in $\mathbb{C}[\boldsymbol{x}]$;
d is the order of deflation;
$\boldsymbol{x}^0 \approx \boldsymbol{x}^*$, $\boldsymbol{x}^* \in F^{-1}(\boldsymbol{0})$, an isolated multiple solution;

Output: $D^{(d)}F$ is a finite set of polynomials in $\mathbb{C}[\boldsymbol{x}, \boldsymbol{\lambda}]$ such that there is $\boldsymbol{\lambda}^*$ with $\mu_{D^{(d)}F}(\boldsymbol{x}^*, \boldsymbol{\lambda}^*) < \mu_F(\boldsymbol{x}^*)$.

determine the numerical corank m of $A^{(d)}(\boldsymbol{x}^0)$;
return $D^{(d)}F := G^{(d)}(\boldsymbol{x}, \boldsymbol{\lambda})$ as in (2.6).

Ideally, it would be nice to have a method to predict a number d such that the system can be regularized by a single deflation of order d. However, at this point, the iterative application of Algorithms 2.2 and 2.3 is the best practical strategy; This gives a procedure that deflates the system completely in at most the number of variables steps, as opposed to the ordinary deflation that may take more steps.

3. Multiplicity structure. This section relates two different ways to obtain the multiplicity of an isolated solution, constructing its *multiplicity structure*. Note that by a "multiplicity structure" – a term without a precise mathematical definition – we mean any structure which provides more local information about the singular solution in addition to its multiplicity. In this section we mention two different approaches to describe this so-called multiplicity structure.

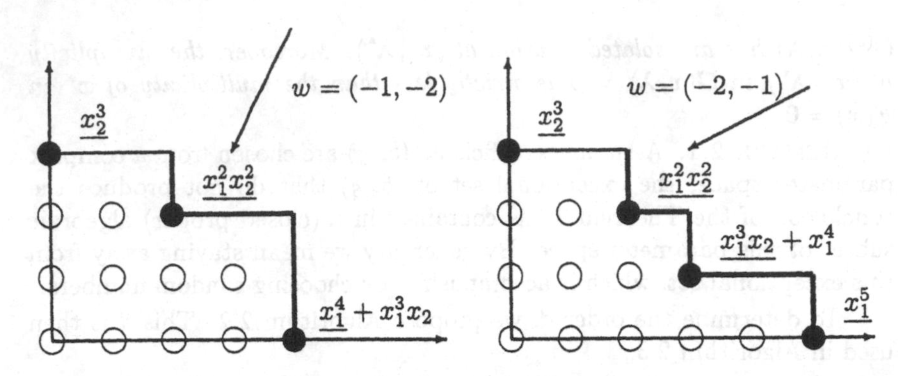

FIG. 1. *Two monomial staircases for two different monomial orderings applied to the same system. The full circles represent the generators of the initial ideals. The multiplicity is the number of standard monomials, represented by the empty circles under the staircase.*

EXAMPLE 2 (Running example 1). Consider the system

$$F(\boldsymbol{x}) = \begin{cases} x_2^3 = 0 \\ x_1^2x_2^2 = 0 \\ x_1^4 + x_1^3x_2 = 0. \end{cases} \tag{3.1}$$

The system $F(\boldsymbol{x}) = \mathbf{0}$ has only one isolated solution at $(0,0)$ of high multiplicity. Below we will show how to compute the multiplicity of $(0,0)$. ◇

3.1. Standard bases. Assume $\mathbf{0} \in \mathbb{C}^n$ is an isolated solution of the system $F(\boldsymbol{x}) = \mathbf{0}$. Let $I = \langle F \rangle \subset R = \mathbb{C}[\boldsymbol{x}]$ be the ideal generated by the polynomials in the system. Given a local monomial order $\geq$, the initial ideal $\text{in}_{\geq}(I) = \{\text{in}_{\geq}(f) \mid f \in I\} \subset R$ describes the multiplicity structure of $\mathbf{0}$ by means of *standard monomials*, i.e.: monomials that are not contained in $\text{in}_{\geq}(I)$. A graphical representation of a monomial ideal is a monomial *staircase*.

EXAMPLE 3 (Initial ideals with respect to a local order). Consider the system (3.1) of Example 2.

Figure 1 shows the staircases for initial ideals of $I = \langle F \rangle$ w.r.t. two local weight orders $\geq_\omega$. Computer algebra systems Macaulay 2 [9] and Singular [12] can be used for these kind of computations, see also [10, 11] for theory, in particular, on Mora's tangent cone algorithm [20].

In the example the leading monomials at the corners of the staircase come from the elements of the corresponding *standard basis.* For the weight vector $w = (-1,-2)$ the original generators give such a basis (initial terms underlined). For $w = (-2,-1)$ one more polynomial is needed. ◇

3.2. Dual space of differential functionals. Another approach at the multiplicity structure is described in detail in [25, 27]; see also [21].

Using duality to define the multiplicity goes back to Macaulay [18]. In this approach, *differential functionals* are denoted by

$$\Delta_\alpha(f) = \frac{1}{\alpha_1!\cdots\alpha_n!}\cdot\frac{\partial^{|\alpha|}f}{\partial x^{\alpha_1}\cdots\partial x^{\alpha_n}}\bigg|_{x=0}. \tag{3.2}$$

Observe that

$$\Delta_\alpha(x^\beta) = \begin{cases} 1, & \alpha=\beta \\ 0, & \alpha\neq\beta. \end{cases} \tag{3.3}$$

We then define the local *dual space* of differential functionals $D_0[I]$ as

$$D_0[I] = \{L \in \text{Span}\{\Delta_\alpha \mid \alpha\in\mathbb{Z}^n_{\geq 0}\} \mid L(f) = 0 \text{ for all } f\in I\}. \tag{3.4}$$

EXAMPLE 4 (Dual space of running example 1). For the ideal defined by the polynomials in the system (3.1) we have

$$D_0[I] = \text{Span}\{\ \begin{matrix} \underline{\Delta_{(4,0)}}-\Delta_{(3,1)}, \underline{\Delta_{(3,0)}}, \underline{\Delta_{(2,1)}}, \underline{\Delta_{(1,2)}}, \\ \underline{\Delta_{(2,0)}}, \underline{\Delta_{(1,1)}}, \underline{\Delta_{(0,2)}}, \underline{\Delta_{(1,0)}}, \underline{\Delta_{(0,1)}}, \underline{\Delta_{(0,0)}} \end{matrix}\ \}. \tag{3.5}$$

Notice that here the basis of the dual space is chosen in such a way that the (underlined) leading terms with respect to the weight order $\geq_{(2,1)}$ correspond to the monomials under the staircase in Example 2 for the order $\geq_{(-2,-1)}$. We will show that it is not a coincidence later in this section. ◇

3.3. Dual bases versus standard bases. Since both local dual bases and initial ideals w.r.t. local orders describe the same, there exists a natural correspondence between the two.

Let $\geq$ be an order on the nonnegative integer lattice $\mathbb{Z}^n_{\geq 0}$ that defines a local monomial order and let $\succeq$ be the opposite of $\geq$: i.e. $\alpha\succeq\beta \Leftrightarrow \alpha\leq\beta$. (Note: $\succeq$ defines a global monomial order.)

For a linear differential functional $L = \sum c_\alpha\Delta_\alpha$ define the *support*: $\text{supp}(L) = \{\alpha\in\mathbb{Z}^n_{\geq 0} \mid c_\alpha\neq 0\}$. For the dual space, $\text{supp}(D_0[I]) = \bigcup_{L\in D_0[I]}\text{supp}(L)$.

Using the order $\succeq$ we can talk about the leading or *initial term* of L: let $\text{in}_\succeq(L)$ be the maximal element of $\text{supp}(L)$ with respect to $\succeq$. Define the *initial support* of the dual space as $\text{in}_\succeq(D_0[I]) = \{\text{in}_\succeq(L) \mid L\in D_0[I]\}$. The initial support is obviously contained in the support, in our running example the containment is proper:

$$\begin{aligned} \text{in}_{(2,1)}(D_0[I]) &= \{(i,j) \mid i+j\leq 3\}\cup\{(4,0)\} \\ &\subset \{(i,j) \mid i+j\leq 3\}\cup\{(4,0)\cup(3,1)\} = \text{supp}(D_0[I]). \end{aligned}$$

THEOREM 3.1. *The number of elements in the initial support equals the dimension of the dual space, therefore, is the multiplicity. Moreover,* *with the above assumptions on the orders $\geq$ and $\succeq$, the standard monomials w.r.t. the local order $\geq$ are* $\{x^\alpha \mid \alpha \in \mathrm{in}_\succeq(D_0[I])\}$.

Proof. Pick $L_\beta \in D_0[I], \beta \in \mathrm{in}_\succeq(D_0[I])$ such that $\mathrm{in}_\succeq(L_\beta) = \beta$. One can easily show that $\{L_\beta\}$ is a basis of $D_0[I]$.

Take a monomial $x^\alpha \in \mathrm{in}_\geq(I)$, then there is $f \in I$ such that $x^\alpha = \mathrm{in}_\geq(f)$. Next, take any linear differential functional L with $\mathrm{in}_\succeq(L) = \alpha$. Since the orders $\geq$ and $\succeq$ are opposite, there are no similar terms in the tail of L and the tail of f, therefore, $L(f) = \mathrm{in}_\succeq(L)(\mathrm{in}_\geq(f)) \neq 0$.

It follows that, $L \notin D_0[I]$, which proves that the set of standard monomials is contained in the initial support of $D_0[I]$. They are equal since they both determine the dimension. □

Consider the ring of linear differential operators $\mathcal{D} = \mathbb{C}[\partial]$ with the natural action (denoted by "$\cdot$") on polynomial ring $R = \mathbb{C}[x]$.

LEMMA 3.1. *Let* $Q \in \mathbb{C}[\partial]$ *and* $f \in \mathbb{C}[x]$ *such that* $\mathrm{in}_\succeq(Q) \succeq \mathrm{in}_\geq(f)$ *(in* $\mathbb{Z}^n_{\geq 0}$*).*

Then $\mathrm{in}_\geq(Q \cdot f) = \mathrm{in}_\geq(f) - \mathrm{in}_\succeq(Q) \in \mathbb{Z}^n_{\geq 0}$.

4. Computing the multiplicity structure. Let the ideal I be generated by $f_1, f_2, \ldots, f_N$. Let $D_0^{(d)}[I]$ the part of $D_0[I]$ containing functionals of order at most d. We would like to have a criterion that for the differential functional L of degree at most d guarantees $L \in D_0^{(d)}[I]$.

Below we describe two such criteria referred to as *closedness conditions*; their names are arranged to match the corresponding computational techniques of Dayton-Zeng [3] and Stetter-Thallinger [26] that we will describe later respectively as the DZ and ST algorithms.

A functional $L = \sum c_\alpha \Delta_\alpha$ with $c_\alpha \in \mathbb{C}$ of order d belongs to the dual space $D_0[I]$ if and only if

- **(DZ-closedness)** $L(g \cdot f_i) = 0$ for all $i = 1, 2, \ldots, N$ and polynomials $g(x)$ of degree at most $d - 1$.
- **(ST-closedness)** $L(f_i) = 0$ for all i and $\sigma_j(L) \in D_0[I]$ for all $j = 1, 2, \ldots, n$, where $\sigma_j : D_0[I] \to D_0[I]$ is a linear map such that

$$\sigma_j(\Delta_\alpha) = \begin{cases} 0, & \text{if } \alpha_j = 0, \\ \Delta_{\alpha - e_j}, & \text{otherwise.} \end{cases} \quad (4.1)$$

The basic idea of both DZ and ST algorithms is the same: build up a basis of D_0 incrementally by computing $D_0^{(d)}$ for $d = 1, 2, \ldots$ using the corresponding closedness condition. The computation stops when $D_0^{(d)} = D_0^{(d-1)}$.

EXAMPLE 5 (Running example 2). Consider the system in $\mathbb{C}[x_1, x_2]$ given by three polynomials $f_1 = x_1 x_2$, $f_2 = x_1^2 - x_2^2$, and $f_3 = x_2^4$, which has only one isolated root at $(0, 0)$. ◇

4.1. The Dayton-Zeng algorithm. We shall outline only a summary of this approach, see [3] for details.

If $\mathbf{0}$ is a solution of the system, then $D_0^{(0)} = \mathrm{Span}\{\Delta_0\}$.

At step $d > 0$, we compute $D_{\mathbf{0}}^{(d)}$. Let the functional

$$L = \sum_{|\alpha| \leq d,\ \alpha \neq \mathbf{0}} c_\alpha \Delta_\alpha \tag{4.2}$$

belong to the dual space $D_{\mathbf{0}}^{(d)}$. Then the vector of coefficients c_α is in the kernel of the following matrix $M_{DZ}^{(d)}$ with $NB(d-1)$ rows and $B(d)-1$ columns, where $B(d) = \binom{n+d}{n}$ is the number of monomials in n variables of degree at most d.

The rows of $M_{DZ}^{(d)}$ are labelled with $x^\alpha f_j$, where $|\alpha| < d$ and $j = 1, 2, \ldots, N$. The columns correspond to Δ_β, where $\beta \neq \mathbf{0}$, $|\beta| \leq d$.

$$[\text{The entry of } M_{DZ}^{(d)} \text{ in row } x^\alpha f_j \text{ and column } \Delta_\beta] = \Delta_\beta(x^\alpha f_j). \tag{4.3}$$

At the step $d = 3$ we have the following $M_{DZ}^{(3)}$

	$\Delta_{(1,0)}$	$\Delta_{(0,1)}$	$\Delta_{(2,0)}$	$\Delta_{(1,1)}$	$\Delta_{(0,2)}$	$\Delta_{(3,0)}$	$\Delta_{(2,1)}$	$\Delta_{(1,2)}$	$\Delta_{(0,3)}$
f_1	0	0	0	1	0	0	0	0	0
f_2	0	0	1	0	-1	0	0	0	0
f_3	0	0	0	0	0	0	0	0	0
$x_1 f_1$	0	0	0	0	0	0	1	0	0
$x_1 f_2$	0	0	0	0	0	1	0	-1	0
$x_1 f_3$	0	0	0	0	0	0	0	0	0
$x_2 f_1$	0	0	0	0	0	0	0	1	0
$x_2 f_2$	0	0	0	0	0	0	1	0	-1
$x_2 f_3$	0	0	0	0	0	0	0	0	0
$x_1^2 f_1$	0	0	0	0	0	0	0	0	0
$\vdots$	$\vdots$	$\vdots$	$\vdots$	$\vdots$	$\vdots$	$\vdots$	$\vdots$	$\vdots$	$\vdots$

Note that the last block of 9 rows is entirely zero.

Analyzing the kernel of this matrix one sees that there are no functionals of degree 3 in the dual space, which is then is equal to $D_{\mathbf{0}}^{(2)}[I]$

$$D_{\mathbf{0}}[I] = \operatorname{Span}\{\Delta_{(0,0)}, \Delta_{(1,0)}, \Delta_{(0,1)}, \Delta_{(2,0)} + \Delta_{(0,2)}\}. \tag{4.4}$$

4.2. The Stetter-Thallinger algorithm. The matrix $M_{ST}^{(d)}$ is a matrix consisting of $n+1$ blocks stacked on top of each other:

- The top block contains the first N rows of $M_{DZ}^{(d)}$;
- For every $j = 1, 2, \ldots, n$, let $S_j^{(d)}$ be the $(B(d-1)-1) \times (B(d)-1)$-matrix for the linear map

$$\sigma_j : D_{\mathbf{0}}^{(d)} / \operatorname{Span}\{\Delta_{\mathbf{0}}\} \to D_{\mathbf{0}}^{(d-1)} / \operatorname{Span}\{\Delta_{\mathbf{0}}\} \tag{4.5}$$

w.r.t. standard bases of functionals.
The block $M_{ST}^{(d-1)} S_j$ represents the closedness condition for the "anti-derivation" σ_j.

Let us go through the steps of the algorithm for the Example 5.

Step 1. At the beginning we have $M_{ST}^{(1)}$ equal to

	$\Delta_{(1,0)}$	$\Delta_{(0,1)}$
f_1	0	0
f_2	0	0
f_3	0	0

Therefore, $D_0^{(1)} = \text{Span}\{\Delta_{(0,0)}, \Delta_{(1,0)}, \Delta_{(0,1)}\}$.

Step 2. Since $M_{ST}^{(1)} S_j^{(2)} = 0$ for all j, the matrix $M_{ST}^{(2)}$ is

	$\Delta_{(1,0)}$	$\Delta_{(0,1)}$	$\Delta_{(2,0)}$	$\Delta_{(1,1)}$	$\Delta_{(0,2)}$
f_1	0	0	0	1	0
f_2	0	0	1	0	-1
f_3	0	0	0	0	0
	0	0	0	0	0
$\vdots$	$\vdots$	$\vdots$	$\vdots$	$\vdots$	$\vdots$

Therefore, $D_0^{(2)} = \text{Span}\{\Delta_{(0,0)}, \Delta_{(1,0)}, \Delta_{(0,1)}, \Delta_{(2,0)} + \Delta_{(0,2)}\}$.

We can "prune" the matrix $M_{ST}^{(2)}$ by row-reducing it to the following matrix with the same kernel:

$$\tilde{M}_{ST}^{(2)} = \begin{bmatrix} 0 & 0 & 0 & 1 & 0 \\ 0 & 0 & 1 & 0 & -1 \end{bmatrix}$$

Step 3. Compute $S_1^{(3)}$ that represents σ_1:

	$\Delta_{(1,0)}$	$\Delta_{(0,1)}$	$\Delta_{(2,0)}$	$\Delta_{(1,1)}$	$\Delta_{(0,2)}$	$\Delta_{(3,0)}$	$\Delta_{(2,1)}$	$\Delta_{(1,2)}$	$\Delta_{(0,3)}$
$\Delta_{(1,0)}$	0	0	1	0	0	0	0	0	0
$\Delta_{(0,1)}$	0	0	0	1	0	0	0	0	0
$\Delta_{(2,0)}$	0	0	0	0	0	1	0	0	0
$\Delta_{(1,1)}$	0	0	0	0	0	0	1	0	0
$\Delta_{(0,2)}$	0	0	0	0	0	0	0	1	0

The matrix $S_2^{(3)}$ can be defined similarly.

The top block of the matrix $M_{ST}^{(3)}$ is

	$\Delta_{(1,0)}$	$\Delta_{(0,1)}$	$\Delta_{(2,0)}$	$\Delta_{(1,1)}$	$\Delta_{(0,2)}$	$\Delta_{(3,0)}$	$\Delta_{(2,1)}$	$\Delta_{(1,2)}$	$\Delta_{(0,3)}$
f_1	0	0	0	1	0	0	0	0	0
f_2	0	0	1	0	-1	0	0	0	0
f_3	0	0	0	0	0	0	0	0	0

Despite the last 4 columns being 0, there are no new elements of order 3 in the dual space due to the other two blocks: $\tilde{M}_{ST}^{(2)} S_1^{(3)}$:

	$\Delta_{(1,0)}$	$\Delta_{(0,1)}$	$\Delta_{(2,0)}$	$\Delta_{(1,1)}$	$\Delta_{(0,2)}$	$\Delta_{(3,0)}$	$\Delta_{(2,1)}$	$\Delta_{(1,2)}$	$\Delta_{(0,3)}$
$x_1 f_1$	0	0	0	0	0	0	1	0	0
$x_1 f_2$	0	0	0	0	0	1	0	−1	0

and $\tilde{M}_{ST}^{(2)} S_2^{(3)}$:

	$\Delta_{(1,0)}$	$\Delta_{(0,1)}$	$\Delta_{(2,0)}$	$\Delta_{(1,1)}$	$\Delta_{(0,2)}$	$\Delta_{(3,0)}$	$\Delta_{(2,1)}$	$\Delta_{(1,2)}$	$\Delta_{(0,3)}$
$x_2 f_1$	0	0	0	0	0	0	0	1	0
$x_2 f_2$	0	0	0	0	0	0	1	0	−1

Comparing to DZ algorithm, in step 3, we managed to avoid the computation of 9 last zero rows of $M_{DZ}^{(3)}$ in this particular example. We now also see how its 4 last nonzero rows show up in the "closedness condition" blocks of $M_{DZ}^{(3)}$.

5. Proofs and algorithmic details. In this section we justify the main theorems stated before and give details about the algorithms presented above.

5.1. First-order deflation. In the beginning of this section we summarize our deflation method introduced in [16]. Not only it is done for the convenience of the reader, but also for our own convenience as we build a higher-order deflation algorithm later in this section following the pattern established for the ordinary deflation.

One deflation step with fixed $\boldsymbol{\lambda}$. The basic idea of the method is relatively simple. Let $\boldsymbol{\lambda} \in \mathbb{C}^n$ be a nonzero vector in $\ker(A(\boldsymbol{x}^*))$, then the equations

$$g_i(\boldsymbol{x}) = \boldsymbol{\lambda} \cdot \nabla f_i(\boldsymbol{x}) = \sum_{i=1}^{n} \lambda_j \frac{\partial f_i(\boldsymbol{x})}{\partial x_j}, \quad i = 1, 2, \ldots, N \tag{5.1}$$

have $\boldsymbol{x}^*$ as a solution. Moreover,

THEOREM 5.1. *The augmented system*

$$G(\boldsymbol{x}) = (f_1, \ldots, f_N, g_1, \ldots, g_N)(\boldsymbol{x}) = \mathbf{0} \tag{5.2}$$

of equations in $\mathbb{C}[\boldsymbol{x}]$ *is a* deflation *of the original system* $F(\boldsymbol{x}) = \mathbf{0}$ *at* x^*, *i.e.* $G(\boldsymbol{x}^*) = 0$ *and the multiplicity of the solution* x^* *is lower in the new system.*

The original proof of this statement in [16] uses the notion of a standard basis of the ideal $I = (f_1, f_2, \ldots, f_N)$ in the polynomial ring $R = \mathbb{C}[\boldsymbol{x}]$ w.r.t. a local order; this tool of computational commutative algebra can be used to obtain the multiplicity of x^*, which is defined as the $\mathbb{C}$-dimension of the local quotient ring R_x/R_xI.

On the other hand it is in correspondence with another way of looking at multiplicities - dual spaces of local functionals, so the proof can be written in that language as well (see Section 3).

One deflation step with indeterminate λ. Without loss of generality, we may assume corank$(A(x^*)) = 1$; consult [16] to see how the general case is reduced to this. Consider $N+1$ additional polynomials in $\mathbb{C}[x, \lambda]$ in $2n$ variables:

$$g_i(x, \lambda) = \lambda \cdot \nabla f_i(x) = \sum_{j=1}^{n} \lambda_j \frac{\partial f_i(x)}{\partial x_j}, \ (i = 1, 2, \ldots, N) \tag{5.3}$$

$$h(\lambda) = \sum_{j=1}^{n} b_j \lambda_j - 1, \tag{5.4}$$

where the coefficients b_j are random complex numbers.

THEOREM 5.2. *Let $x^* \in \mathbb{C}^n$ be an isolated solution of $F(x) = 0$ (in $\mathbb{C}[x]$).*

For a generic choice of coefficients b_j, $j = 1, 2, \ldots, n$, there exists a unique $\lambda^* \in \mathbb{C}^n$ such that the system

$$G(x, \lambda) = (f_1, \ldots, f_N, g_1, \ldots, g_N, h)(x, \lambda) = 0 \tag{5.5}$$

of equations in $\mathbb{C}[x, \lambda]$ has an isolated solution at (x^*, λ^*).

The multiplicity of (x^*, λ^*) in $G(x, \lambda) = 0$ is lower than that of x^* in $F(x) = 0$.

Proof. Follows from Proposition 3.4 in [16]. □

Theorem 5.2 provides a recipe for the deflation algorithm: one simply needs to keep deflating until the solution of the augmented system corresponding to x^* becomes regular.

As a corollary we have that the number of deflations needed to make a singular isolated solution x^* regular is less than the multiplicity of x^*.

5.2. Higher-order deflation with fixed multipliers. We use the deflation operator to define an augmented system.

THEOREM 5.3. *Let $f_1, f_2, \ldots, f_N$ form a standard basis of I w.r.t. the order opposite to $\succeq$. Consider the system $G^{(d)}(x) = 0$ in $\mathbb{C}[x]$, where*

$$G^{(d)}(x) = \begin{cases} f_j(x) & (j = 1, 2, \ldots, N) \\ g_{j,\alpha}(x) & (j = 1, 2, \ldots, N, \ |\alpha| < d) \end{cases} . \tag{5.6}$$

as in Definition 2.2.

(a) The system $G^{(d)}(x) = 0$ is a deflation of the original system $F(x) = 0$ at x^*.

(b) *Let $I = (F)$ and $J = (G^{(d)})$ be the ideals generated by polynomials of the systems and $\succeq$ be a global monomial order on $\mathbb{Z}^n_{\geq 0}$. Then the following relation holds for initial supports*

$$\mathrm{in}_{\succeq}(D_0[J]) \subset \{\beta - \beta_Q \mid \beta \in \mathrm{in}_{\succeq}(D_0[I])\} \cap \mathbb{Z}^n_{\geq 0}, \qquad (5.7)$$

where β_Q is the maximal element of the set $\mathrm{in}_{\succeq}(D_0[I]) \cap \{\beta : |\beta| \leq d\}$.

Proof. Let $\boldsymbol{\lambda} \in \ker(A^{(d)}(\boldsymbol{x}^*))$ be the vector used above to construct the operator $Q \in \mathbb{C}[\partial]$ and the equations $g_{j,\alpha}(\boldsymbol{x}) = 0$.

First of all, $g_{j,\alpha}(\boldsymbol{x}^*) = (Q \cdot (\boldsymbol{x}^\alpha f_j))|_{\boldsymbol{x}=\boldsymbol{x}^*} = 0$ provided $|\alpha| < d$ (by construction), hence, $\boldsymbol{x}^*$ is a solution to $G^{(d)}(\boldsymbol{x}) = \mathbf{0}$.

To prove (a), it remains to show that the multiplicity drops, which follows from part (b) that is treated in the rest of this proof.

We shall assume for simplicity that $\boldsymbol{x}^* = \mathbf{0}$. This is done without the loss of generality using a linear change of coordinates: $\boldsymbol{x} \mapsto \boldsymbol{x} + \boldsymbol{x}^*$. It is important to note that in the new coordinates polynomials $Q \cdot (\boldsymbol{x}^\alpha f_j(\boldsymbol{x} + \boldsymbol{x}^*))$ generate the same ideal as the polynomials $Q \cdot ((\boldsymbol{x} - \boldsymbol{x}^*)^\alpha f_j(\boldsymbol{x} + \boldsymbol{x}^*))$.

Recall that $I = \langle F \rangle = \langle f_1, f_2, \ldots, f_N \rangle$, let $J = \langle G^{(d)} \rangle \supset I$ be the ideal generated by the polynomials in the augmented system. The reversed containment holds for the dual spaces: $D_0[I] \supset D_0[J]$.

There is a 1-to-1 correspondence between linear differential operators and linear differential functionals:

$$\sum \lambda_\beta \partial^\beta \longleftrightarrow \sum \lambda_\beta \beta! \Delta_\beta. \qquad (5.8)$$

Let $\phi : \mathbb{C}[\partial] \to D_0$ and $\tau : D_0 \to \mathbb{C}[\partial]$ be the corresponding bijections.

As in Section 3 we order terms Δ_β with $\succeq$, a global monomial order. Notice that since the choice of coefficients of the operator Q is generic, $\beta_Q = \mathrm{in}_{\succeq}(Q) = \mathrm{in}_{\succeq}(\phi(Q))$ is the maximal element of the set $\mathrm{in}_{\succeq}(D_0[I]) \cap \{\beta : |\beta| \leq d\}$.

Next, we use the condition that f_i form a standard basis. Since the corners of the staircase correspond to the initial terms of f_i, by Lemma 3.1 the staircase created with the corners at $\mathrm{in}_{\geq}(Q \cdot (\boldsymbol{x}^\alpha f_i))$ bounds the set $\{\beta - \beta_Q \mid \beta \in \mathrm{in}_{\succeq}(D_0[I])\} \cap \mathbb{Z}^n_{\geq 0}$, which, therefore, contains the initial support of $D_0[J]$. □

COROLLARY 5.1. *If there exist a local monomial order $\geq$ such that the minimal (standard) monomial in the set $\{\boldsymbol{x}^\alpha \notin \mathrm{in}_{\geq}(I) : |\alpha| \leq d\}$ is also minimal in the set of all standard monomials, i.e., $\{\boldsymbol{x}^\alpha \notin \mathrm{in}_{\geq}(I)\}$, then $\boldsymbol{x}^*$ is a regular solution of $G^{(d)}(\boldsymbol{x}) = \mathbf{0}$.*

Theorem 5.3 and Corollary 5.1 are of purely theoretical nature, since the assumption of exactness in their statements can not be relaxed.

Ideally, we would like to be able to drop the assumption of the original polynomials forming a standard basis, since computing such a basis is a complex symbolic task, whereas our interest lies in the further numericalization of the approach. The following weaker statement works around this restriction.

Assuming $x^* = \mathbf{0}$, let $\operatorname{supp}(F) = \bigcup_{j=1,2,\dots,N} \operatorname{supp}(f_j)$.

PROPOSITION 5.1. *Assume $A(\mathbf{0}) = \mathbf{0}$. Let $d_0 = \min\{|\alpha| : x^\alpha \in \operatorname{supp}(F)\}$.*

Then, in the notation of Theorem 5.3, for a generic deflating operator Q the system $G^{(d)}(x) = \mathbf{0}$, where, $d < d_0$ is a deflation of the original system $F(x) = \mathbf{0}$ at the origin.

Moreover, if $d = d_0 - 1$ then the Jacobian of $G^{(d)}(\mathbf{0})$ is not equal to zero.

Proof. Fix a local monomial ordering that respects the degree. With the above assumptions, the initial ideal $\operatorname{in}(\langle F \rangle)$ will contain monomials of degree at least d_0. On the other hand, for a generic choice of the deflating operator Q the support $\operatorname{supp}(G^{(d)})$ would contain a monomial of degree less than d_0. Therefore, there exists a monomial in $\operatorname{in}(\langle \operatorname{supp}(G^{(d)}) \rangle)$ that is not in $\operatorname{in}(\langle F \rangle)$, hence, $G^{(d)}$ is a deflation.

If $d = |d_0| - 1$, then there is such monomial of degree 1, which means that the Jacobian of the augmented system is nonzero. □

REMARK 5.1. Note that if the deflation order d is as in Proposition 5.1, then it suffices to take an arbitrary *homogeneous* deflation operator of order d.

Next we explain the practical value of Proposition 5.1. Let $K = \ker A(\mathbf{0})$ and $c = \operatorname{corank} A(\mathbf{0}) = \dim K$. Without a loss of generality we may assume that K is the subspace of $\mathbb{C}^n$ has $\{x_1, \dots, x_c\}$ as coordinates.

Now consider the system $F'(x_1, \dots, x_c) = F(x_1, \dots, x_c, 0, \dots, 0)$. This system has an isolated solution at the origin, and Proposition 5.1 is applicable, since the Jacobian is zero. Moreover, if we take the deflation of order $d = d_0 - 1$ of the original system F, with d_0 coming from the Proposition, the corank of the Jacobian the augmented system $G^{(d)}$ is guaranteed to be lower than that of $A(\mathbf{0})$.

Let us go back to the general setup: an arbitrary isolated solution x^*, the Jacobian $A(x^*)$ with a proper kernel K, etc. Algorithm 2.2 is a practical algorithm that can be executed numerically knowing only an approximation to x^*.

Proof [Proof of correctness of Algorithm 2.2 for $x^0 = x^*$ and $\varepsilon = 0$]. We can get to the special setting of Proposition 5.1 in two steps. First, apply an affine transformation that takes x^* to the origin and $\ker A(x^*)$ to the subspace K of $\mathbb{C}^n$ spanned by the first $c = \operatorname{corank} A(x^*)$ standard basis vectors. Second, make a new system $F'(x_1, \dots, x_c) = 0$ by substituting the $x_i = 0$ in F for $i > c$.

Let $\gamma' = (\gamma'_1, \dots, \gamma'_c) \in K$ be the image of the generic vector γ under the linear part of the affine transform. Then $H(t) = F'(\gamma'_1 t, \dots, \gamma'_c t)$.

Since γ' is generic, the lowest degree d_0 of the monomial in $\operatorname{supp}(F')$ is equal to $\min\{a \mid t^a \in \operatorname{supp} H(t)\}$. According to the Proposition 5.1 and the discussion that followed, $d = d_0 - 1$ is the minimal order of deflation that will reduce the rank of the system. □

REMARK 5.2. In view of Remark 5.1 it would be enough to use any *homogeneous* deflation operator of order d:

$$Q = \sum_{|\beta|=d} \lambda_\beta \partial^\beta \quad \in \mathbb{C}[\partial], \tag{5.9}$$

such that the vector λ of its coefficients is in the kernel of the *truncated deflation matrix*, which contains only the rows corresponding to the original polynomials F and only the columns labelled with ∂^β with $|\beta| = d$.

5.3. Indeterminate multipliers. As in Section 5.1, we now consider indeterminate λ_β. Now we should think of the differential operator $L(\lambda) \in \mathbb{C}[\lambda, \partial]$ and of additional equations $g_{j,\alpha}(x, \lambda) \in \mathbb{C}[x, \lambda]$ as depending on λ.

Proof of Theorem 2.1. Picking $m = \operatorname{corank}(A^{(d)}(x^*))$ generic linear equations h_k guarantees that for $x = x^*$ the solution for λ exists and is unique; therefore, the first part of the statement is proved.

The argument for the drop in the multiplicity is similar to that of the proof of Theorem 5.2. □

6. Computational experiments. We have implemented our new deflation methods in PHCpack [28] and Maple. Below we report on two examples.

One crucial decision in the deflation algorithm is the determination of the numerical rank, for which we may use SVD or QR in the rank-revealing algorithms. Both SVD and QR are numerically stable, As shown by the result from [4, page 118] for the problem of solving an overdetermined linear system $Ax = b$. The solution obtained by QR or SVD minimizes the residual $||(A + \delta A)\tilde{x} - (b + \delta b)||_2$ where the relative errors have the same magnitude as the machine precision ϵ:

$$\max\left(\frac{||\delta A||_2}{||A||_2}, \frac{||\delta b||_2}{||b||_2}\right) = O(\epsilon). \tag{6.1}$$

To decide to tolerance for the numerical rank, we combine a fixed value – relative to the working precision – with the search for that value of i where the maximal jump σ_{i+1}/σ_i in the singular values σ_i's occurs.

6.1. A first example. To find initial approximations for the roots of the system

$$F(x) = \begin{cases} x_1^3 + x_1 x_2^2 = 0 \\ x_1 x_2^2 + x_2^3 = 0 \\ x_1^2 x_2 + x_1 x_2^2 = 0 \end{cases} \tag{6.2}$$

we must first make the system "square", i.e.: having as many equations as unknowns, so we may apply the homotopies available in PHCpack [28]. Using the embedding technique of [23] (see also [24]), we add one slack variable z to each equation of the system, multiplied by random complex constants γ_1, γ_2, and γ_3:

$$E(\boldsymbol{x}, z) = \begin{cases} x_1^3 + x_1 x_2^2 + \gamma_1 z = 0 \\ x_1 x_2^2 + x_2^3 + \gamma_2 z = 0 \\ x_1^2 x_2 + x_1 x_2^2 + \gamma_3 z = 0. \end{cases} \tag{6.3}$$

Observe that the solutions of the original system $F(\boldsymbol{x}) = \mathbf{0}$ occur as solutions of the embedded system $E(\boldsymbol{x}, z) = \mathbf{0}$ with slack variable $z = 0$. At the end points of the solution paths defined by a homotopy to solve $E(\boldsymbol{x}, z) = \mathbf{0}$, we find nine zeroes close to the origin. These nine approximate zeroes are the input to our deflation algorithm.

The application of our first deflation algorithm in [16] requires two stages. The Jacobian matrix of $F(\boldsymbol{x}) = \mathbf{0}$ has rank zero at $(0,0)$. After the first deflation with one multiplier, the rank of the Jacobian matrix of the augmented system $G(\boldsymbol{x}, \boldsymbol{\lambda}_1) = \mathbf{0}$ equals one, so the second deflation step uses two multipliers. After the second deflation step, the Jacobian matrix has full rank, and $(0,0)$ has then become a regular solution. Newton's method on the final system then converges again quadratically and the solution can be approximated efficiently with great accuracy. Once the precise location of a multiple root is known, we are interested in its multiplicity. The algorithm of [3] reveals that the multiplicity of the isolated root equals seven.

Starting at a root of low accuracy, at a distance of 10^{-5} from the exact root, the numerical implementation of Algorithm 2.2 predicts two as the order, using 10^{-4} as the tolerance for the vanishing of the coefficients in the univariate interpolating polynomial. The Jacobian matrix of the augmented system $G^{(2)}$ has full rank so that a couple of iterations suffice to compute the root very accurately.

6.2. A larger example. The following system is copied from [14]:

$$F(\boldsymbol{x}) = \begin{cases} 2x_1 + 2x_1^2 + 2x_2 + 2x_2^2 + x_3^2 - 1 = 0 \\ (x_1 + x_2 - x_3 - 1)^3 - x_1^3 = 0 \\ (2x_1^3 + 2x_2^2 + 10x_3 + 5x_3^2 + 5)^3 - 1000x_1^5 = 0. \end{cases} \tag{6.4}$$

Counted with multiplicities, the system has 54 isolated solutions. We focus on the solution $(0, 0, -1)$ which occurs with multiplicity 18.

Although Algorithm 1 suggests that the first-order deflation would already lower the corank of the system, we would like to search for a homogeneous deflation operator Q of order two.

To this end we construct the (truncated) deflation matrix $\bar{A}(x_1, x_2, x_3)$ which corresponds to $\{\partial_1^2, \partial_1\partial_2, \partial_1\partial_3, \partial_2^2, \partial_2\partial_3, \partial_3^2\}$, having 12 rows and only 6 columns.

The vectors $(1,6,8,-3,0,4)^T$ and $(0,3,3,-1,1,2)^T$ span the kernel of $\bar{A}(0,0,-1)$. The operator corresponding to the former,

$$Q = \partial_1^2 + 6\partial_1\partial_2 + 8\partial_1\partial_3 - 3\partial_2^2 + 4\partial_3^2, \tag{6.5}$$

regularizes the system, since the equations

$$\begin{cases} Q\cdot(x_1f_1) &= 8x_1 + 24x_2 + 16x_3 + 16 = 0 \\ Q\cdot(x_2f_1) &= 24x_1 - 24x_2 = 0 \\ Q\cdot(x_3f_1) &= 32x_1 + 16x_3 + 16 = 0. \end{cases} \tag{6.6}$$

augmented to the original equations, give a system with the full-rank Jacobian matrix at $(0,0,-1)$.

7. Conclusion. In this paper we have described two methods of computing the multiplicity structure at isolated solutions of polynomial systems. We have developed a higher-order deflation algorithm that reduces the multiplicity faster than the first-order deflation in [16].

In our opinion, one of the main benefits of the higher order deflation for the numerical algebraic geometry algorithms is the possibility to regularize the system in a single step. For that one has to determine the minimal order of such a deflation or, even better, construct a sparse ansatz for its deflation operator. Predicting these numerically could be a very challenging task, which should be explored in the future.

Acknowledgements. The first author would like to thank RICAM/RISC for supporting his visit during the special semester on Gröbner bases. The first two authors received support from the IMA during the thematic year on Applications of Algebraic Geometry. We also thank the referees for useful comments.

REFERENCES

[1] E.L Allgower, K. Böhmer, A. Hoy, and V. Janovský. Direct methods for solving singular nonlinear equations. *ZAMM Z. Angew. Math. Meth.*, **79**(4): 219–231, 1999.

[2] D.J. Bates, C. Peterson, and A.J. Sommese. A numerical-symbolic algorithm for computing the multiplicity of a component of an algebraic set. *J. Complexity*, **22**(4): 475–489, 2006.

[3] B.H. Dayton and Z. Zeng. Computing the multiplicity structure in solving polynomial systems. In M. Kauers, editor, *Proceedings of the 2005 International Symposium on Symbolic and Algebraic Computation*, pp. 116–123. ACM, 2005.

[4] J.W. Demmel. *Applied Numerical Linear Algebra*, Vol. **45** of *Classics in Applied Mathematics*. SIAM, 2003.

[5] P. Deuflhard. *Newton Methods for Nonlinear Problems. Affine Invariance and Adaptive Algorithms.* Springer-Verlag, 2004.

[6] M. Giusti, G. Lecerf, B. Salvy, and J.C. Yakoubsohn. On location and approximation of clusters of zeroes of analytic functions. *Foundations of Computational Mathematics* **5**(3): 257-311, 2005.

[7] M. Giusti, G. Lecerf, B. Salvy, and J.C. Yakoubsohn. On location and approximation of clusters of zeroes: Case of embedding dimension one. *Foundations of Computational Mathematics* **7**(1): 1–58, 2007.

[8] W.J.F. Govaerts. *Numerical Methods for Bifurcations of Dynamical Equilibria.* SIAM, 2000.

[9] D. Grayson and M. Stillman. Macaulay 2, a software system for research in algebraic geometry. Available at http://www.math.uiuc.edu/Macaulay2/.

[10] G.-M. Greuel and G. Pfister. Advances and improvements in the theory of standard bases and syzygies. *Arch. Math.*, **66**: 163–196, 1996.

[11] G.-M. Greuel and G. Pfister. *A Singular Introduction to Commutative Algebra.* Springer-Verlag, 2002.

[12] G.-M. Greuel, G. Pfister, and H. Schönemann. Singular 2.0. A Computer Algebra System for Polynomial Computations, Centre for Computer Algebra, University of Kaiserslautern, 2001. `http://www.singular.uni-kl.de`.

[13] P. Kunkel. A tree-based analysis of a family of augmented systems for the computation of singular points. *IMA J. Numer. Anal.*, **16**: 501–527, 1996.

[14] G. Lecerf. Quadratic Newton iteration for systems with multiplicity. *Found. Comput. Math.*, **2**: 247–293, 2002.

[15] A. Leykin, J. Verschelde, and A. Zhao. Evaluation of Jacobian matrices for Newton's method with deflation to approximate isolated singular solutions of polynomial systems. In *Symbolic-Numeric Computation*, edited by D. Wang and L. Zhi, pp. 269–278. Trends in Mathematics, Birkhäuser, 2007.

[16] A. Leykin, J. Verschelde, and A. Zhao. Newton's method with deflation for isolated singularities of polynomial systems. *Theoretical Computer Science*, **359**(1–3): 111–122, 2006.

[17] Y. Lijun. On the generalized Lyapunov-Schmidt reduction. *ZAMM Z. Angew. Math. Mech.*, **84**(8): 528–537, 2004.

[18] F.S. Macaulay. *The Algebraic Theory of Modular Systems.* Cambridge University Press, 1916. Reissued with an Introduction by Paul Roberts in the Cambridge Mathematical Library 1994.

[19] Z. Mei. *Numerical Bifurcation Analysis for Reaction-Diffusion Equations*, Vol. **28** of Springer Series in Computational Mathematics, Springer-Verlag, 2000.

[20] F. Mora. An algorithm to compute the equations of tangent cones. In J. Calmet, editor, *Computer Algebra. EUROCAM'82, European Computer Algebra Conference. Marseille, France, April 1982*, Vol. **144** of *Lecture Notes in Computer Science*, pp. 158–165. Springer-Verlag, 1982.

[21] B. Mourrain. Isolated points, duality and residues. *Journal of Pure and Applied Algebra*, **117/118**: 469–493, 1997.

[22] T. Ojika, S. Watanabe, and T. Mitsui. Deflation algorithm for the multiple roots of a system of nonlinear equations. *J. Math. Anal. Appl.*, **96**: 463–479, 1983.

[23] A.J. Sommese and J. Verschelde. Numerical homotopies to compute generic points on positive dimensional algebraic sets. *J. of Complexity*, **16**(3): 572–602, 2000.

[24] A.J. Sommese and C.W. Wampler. *The Numerical solution of systems of polynomials arising in engineering and science.* World Scientific, 2005.

[25] H.J. Stetter. *Numerical Polynomial Algebra.* SIAM, 2004.

[26] H.J. Stetter and G.T. Thallinger. Singular systems of polynomials. In O. Gloor, editor, *Proceedings of the 1998 International Symposium on Symbolic and Algebraic Computation*, pp. 9–16. ACM, 1998.

[27] G.T. Thallinger. *Zero Behavior in Perturbed Systems of Polynomial Equations.* PhD thesis, Tech. Univ. Vienna, 1998.

[28] J. Verschelde. Algorithm 795: PHCpack: A general-purpose solver for polynomial systems by homotopy continuation. *ACM Trans. Math. Softw.*, **25**(2): 251–276, 1999. Software available at `http://www.math.uic.edu/~jan`.

[29] J. Verschelde and A. Zhao. Newton's method with deflation for isolated singularities. Poster presented at ISSAC'04, 6 July 2004, Santander, Spain. Available at `http://www.math.uic.edu/~jan/poster.pdf` and at `http://www.math.uic.edu/~azhao1/poster.pdf`.

POLARS OF REAL SINGULAR PLANE CURVES

HEIDI CAMILLA MORK* AND RAGNI PIENE*

Abstract. Polar varieties have in recent years been used by Bank, Giusti, Heintz, Mbakop, and Pardo, and by Safey El Din and Schost, to find efficient procedures for determining points on all real components of a given non-singular algebraic variety. In this note we review the classical notion of polars and polar varieties, as well as the construction of what we here call reciprocal polar varieties. In particular we consider the case of real affine plane curves, and we give conditions for when the polar varieties of singular curves contain points on all real components.

Key words. Polar varieties, hypersurfaces, plane curves, tangent space, flag.

AMS(MOS) subject classifications. Primary 14H50; Secondary 14J70,14P05, 14Q05.

1. Introduction. The general study of polar varieties goes back to Severi, though polars of plane curves were introduced already by Poncelet. Polar varieties of complex, projective, non-singular varieties have later been studied by many, and the theory has been extended to singular varieties (see [6] and references therein). We shall refer to these polar varieties as *classical* polar varieties.

Another kind of polar varieties, which we here will call *reciprocal* polar varieties, were introduced in [3] under the name of dual polar varieties. The definition involves a quadric hypersurface and polarity of linear spaces with respect to this quadric. Classically, the *reciprocal curve* of a plane curve was defined to be the curve consisting of the polar points of the tangent lines of the curve with respect to a given conic. The reciprocal curve is isomorphic to the dual curve in $(\mathbb{P}^2)^\vee$, via the isomorphism of $\mathbb{P}^2$ and $(\mathbb{P}^2)^\vee$ given by the quadratic form defining the conic.

Bank, Giusti, Heintz, Mbakop, and Pardo have proved [1–4] that polar varieties of *real, affine* non-singular varieties (with some requirements) contain points on each connected component of the variety, and this property is useful in CAGD for finding a point on each component. Related work has been done by Safey El Din and Schost [7, 8].

We will in this paper determine in which cases the polar varieties of a real affine *singular* plane curve contain at least one non-singular point of each component of the curve. In the next section we briefly review the definitions of classical and reciprocal polar varieties, and state and partly prove versions of some results from [3] adapted to our situation. The third section treats the case of plane curves; we show that the presence of ordinary multiple points does not affect these results, but that the presence of arbitrary singularities does.

*CMA, Department of Mathematics, University of Oslo, P.O. Box 1053 Blindern, NO-0316 Oslo, Norway ({heidisu,ragnip}@math.uio.no).

2. Polar varieties. Let $V \subset \mathbb{P}^n$ be a complex projective variety. Given a hyperplane H we can consider the affine space $\mathbb{A}^n := \mathbb{P}^n \setminus H$, where H is called the hyperplane at infinity. We define the corresponding affine variety S to be the variety $V \cap \mathbb{A}^n$. We let $V_{\mathbb{R}}$ and $S_{\mathbb{R}}$ denote the corresponding real varieties.

If L_1 and L_2 are linear varieties in a projective space $\mathbb{P}^n$, we let $\langle L_1, L_2\rangle$ denote the linear variety spanned by them. We say that L_1 and L_2 intersect transversally if $\langle L_1, L_2\rangle = \mathbb{P}^n$; if they do not intersect transversally, we write $L_1 \not\pitchfork L_2$.

If I_1 and I_2 are sub-vector spaces of $\mathbb{A}^n$ (considered as a vector space), we say that I_1 and I_2 intersect transversally if $I_1 + I_2 = \mathbb{A}^n$; if they do not intersect transversally, we write $I_1 \not\pitchfork I_2$.

2.1. Classical polar varieties. Let us recall the definition of the classical polar varieties (or loci) of a possibly singular variety. Consider a flag of linear varieties in $\mathbb{P}^n$,

$$\mathcal{L} : L_0 \subset L_1 \subset \ldots \subset L_{n-1} \subset \mathbb{P}^n$$

and a (non degenerate) variety $V \subset \mathbb{P}^n$ of codimension p. Let V_{ns} denote the set of nonsingular points of V. For each point $P \in V_{\mathrm{ns}}$, let $T_P V$ denote the *projective* tangent space to V at P. The i-th polar variety $W_{L_{i+p-2}}(V)$, $1 \le i \le n-p$, of V with respect to $\mathcal{L}$ is the Zariski closure of the set

$$\{P \in V_{\mathrm{ns}} \setminus L_{i+p-2} \,|\, T_P V \not\pitchfork L_{i+p-2}\}.$$

Take $H = L_{n-1}$ to be the hyperplane at infinity and let $S = V \cap \mathbb{A}^n$ be the affine part of V. Then we can define the *affine* polar varieties of S with respect to the flag $\mathcal{L}$ as follows: the i-th affine polar variety $W_{L_{i+p-2}}(S)$ of S with respect to $\mathcal{L}$ is the intersection $W_{L_{i+p-2}}(V) \cap \mathbb{A}^n$.

Since L_{n-1} is the hyperplane at infinity, and all the other elements of the flag $\mathcal{L}$ are contained in L_{n-1}, we can look at the affine cone over each element L_j of the flag, considered as a $(j+1)$-dimensional linear sub-vector space I_{j+1} of $\mathbb{A}^n$. Hence we get the flag (of vector spaces)

$$\mathcal{I} : I_1 \subset I_2 \subset \ldots \subset I_{n-1} \subset \mathbb{A}^n,$$

and the affine polar variety $W_{L_{j-1}}(S)$ can also be defined as the closure of the set

$$\{P \in S_{\mathrm{ns}} \,|\, t_P S \not\pitchfork I_j\},$$

where $t_P S$ denotes the *affine* tangent space to S at P translated to the origin (hence considered as a sub-vector space of $\mathbb{A}^n$). For more details on these two equivalent definitions of affine polar varieties, see [3].

If $f_1, \ldots, f_r$ are homogeneous polynomials in $n+1$ variables, we let $\mathcal{V}(f_1, \ldots, f_r) \subset \mathbb{P}^n$ denote the corresponding algebraic variety (their common zero set). Similarly, if $f_1, \ldots, f_r$ are polynomials in n variables, we

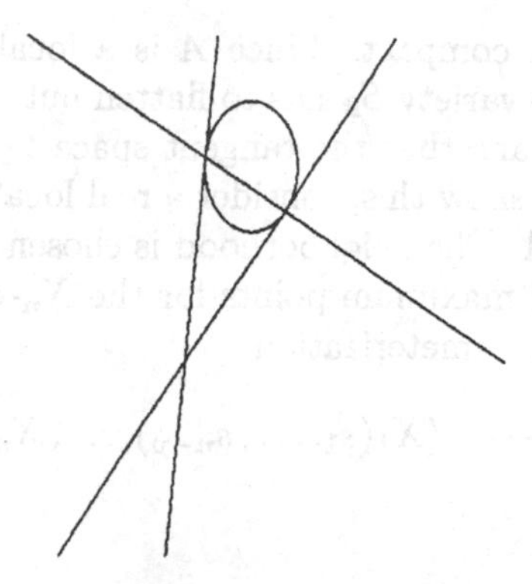

FIG. 1. *A conic with two tangents. The polar locus of the conic (with respect to the point of intersection of the two tangents) consists of the two points of tangency. The* polar *of the point is the line through the two points of tangency.*

denote by $\mathcal{V}(f_1, \ldots, f_r) \subset \mathbb{A}^n$ the corresponding affine variety. We shall often use the flag

$$\mathcal{L} : \mathcal{V}(X_0, X_2, \ldots, X_n) \subset \ldots \subset \mathcal{V}(X_0, X_n) \subset \mathcal{V}(X_0) \subset \mathbb{P}^n \tag{2.1}$$

and the corresponding affine flag (considered as vector spaces)

$$\mathcal{I} : \mathcal{V}(X_2, \ldots, X_n) \subset \ldots \subset \mathcal{V}(X_{n-1}, X_n) \subset \mathcal{V}(X_n) \subset \mathbb{A}^n. \tag{2.2}$$

The following proposition is part of a proposition stated and proved in [2, 2.4, p. 134]. Since this particular result is valid under fewer assumptions, we shall give a different and more intuitive proof. We let L denote any member of the flag $\mathcal{L}$.

PROPOSITION 2.1. *Let $S \subset \mathbb{A}^n$ be a pure p-codimensional reduced variety. Suppose that $S_{\mathbb{R}}$ is non-empty, pure p-codimensional, non-singular, and compact. Then $W_L(S_{\mathbb{R}})$ contains at least one point of each connected component of $S_{\mathbb{R}}$.*

Proof. By an affine coordinate change, we may assume that the flag $\mathcal{I}$ is the flag (2.2). The polar varieties form a sequence of inclusions

$$W_{L_{n-2}}(S_{\mathbb{R}}) \subset W_{L_{n-3}}(S_{\mathbb{R}}) \subset \ldots \subset W_{L_{p-1}}(S_{\mathbb{R}}),$$

so it is sufficient to find a point of $W_{L_{n-2}}(S_{\mathbb{R}})$ on each component of the variety $S_{\mathbb{R}}$. We will show that the maximum point for the last coordinate X_n of each component of $S_{\mathbb{R}}$ is also a point on $W_{L_{n-2}}(S_{\mathbb{R}})$.

The variety $W_{L_{n-2}}(S_{\mathbb{R}})$ is the set

$$\{P \in S_{\mathbb{R}} \mid t_P S \not\pitchfork \mathcal{V}(X_n)\}.$$

Let C be any connected component of $S_{\mathbb{R}}$, and let $A := (a_1, \ldots, a_n) \in C$ be a local maximum point for the last coordinate X_n in C. Such a maximum point exists since C is compact. Since A is a local maximum point for the X_n-coordinate, the variety $S_{\mathbb{R}}$ has to flatten out in the X_n-direction in the point A, which means that the tangent space $t_A S$ is contained in the hyperplane $\mathcal{V}(X_n)$. To show this, consider a real local parameterization $S_{\mathbb{R}}$ in a neighborhood of A. The neighborhood is chosen to be so small that it contains no other local maximum points for the X_n-coordinate.

Consider a local parameterization

$$(s_1, \ldots, s_{n-p}) \longmapsto (X_1(s_1, \ldots, s_{n-p}), \ldots, X_n(s_1, \ldots, s_{n-p})),$$

with

$$A = (X_1(0, \ldots, 0), \ldots, X_n(0, \ldots, 0)).$$

The rows of the matrix

$$\begin{bmatrix} \frac{\partial X_1}{\partial s_1}(0, \ldots, 0) & \cdots & \frac{\partial X_n}{\partial s_1}(0, \ldots, 0) \\ \vdots & \ddots & \vdots \\ \frac{\partial X_1}{\partial s_{n-p}}(0, \ldots, 0) & \cdots & \frac{\partial X_n}{\partial s_{n-p}}(0, \ldots, 0) \end{bmatrix}$$

span the tangent space $t_A S$ of $S_{\mathbb{R}}$ at A. We will show that $\frac{\partial X_n}{\partial s_i}(0, \ldots, 0) = 0$ for $i = 1, \ldots, n-p$. Then we know that $t_A S$ is contained in the hyperplane $\mathcal{V}(X_n)$.

By the definition of derivatives we have that

$$\begin{aligned} \frac{\partial X_n}{\partial s_i}(0, \ldots, 0) &= \lim_{s_i \to 0^-} \frac{X_n(0,\ldots,0,s_i,0,\ldots,0) - X_n(0,\ldots,0)}{s_i - 0} \\ &= \lim_{s_i \to 0^+} \frac{X_n(0,\ldots,0,s_i,0,\ldots,0) - X_n(0,\ldots,0)}{s_i - 0} \end{aligned}$$

If s_i goes to 0 from below, then s_i is negative, and $X_n(0, \ldots, 0, s_i, 0, \ldots, 0) - X_n(0, \ldots, 0) \leq 0$ since $X_n(0, \ldots, 0)$ is the local maximum, so we have

$$\lim_{s_i \to 0^-} \frac{X_n(0,\ldots,0,s_i,0,\ldots,0) - X_n(0,\ldots,0)}{s_i - 0} \geq 0$$

When s_i goes to 0 from above, then s_i is positive, so

$$\lim_{s_i \to 0^+} \frac{X_n(0,\ldots,0,s_i,0,\ldots,0) - X_n(0,\ldots,0)}{s_i - 0} \leq 0$$

Since these limits are equal, they both have to be zero, so $\frac{\partial X_n}{\partial s_i}(0, \ldots, 0) = 0$ for $i = 1, \ldots, n-p$.

Since $t_A S$ is contained in $\mathcal{V}(X_n)$, we have $t_A S \not\pitchfork \mathcal{V}(X_n)$, and hence A is a point of $W_{L_{n-2}}(S_{\mathbb{R}})$. □

In the proof of this proposition we can replace the phrase "local maximum point of the X_n-coordinate" with "local minimum point ...", since the consideration we did for the tangent space is the same in both cases. So we have found two points of each connected component, unless the local maximum and the local minimum for the last coordinate coincide. The local maximum and the local minimum coincide if and only if the variety is contained in the hyperplane $\mathcal{V}(X_n)$, which implies that V is degenerate.

In the above proposition we did not assume that the coordinates were in generic position with respect to the polynomials generating the variety, so we allow situations where the polar variety can contain a piece of a component of dimension greater than zero.

Safey El Din and Schost [8] show that one can find points on each component of a smooth, not necessarily compact, affine real variety. They use all the polar varieties given by a flag, and intersect each of these varieties with other varieties. The union of these intersections will be zero-dimensional and it contains points from each connected component of the variety we started with.

2.2. Reciprocal polar varieties. Let $Q = \mathcal{V}(q)$ be a non-degenerate hyperquadric defined in $\mathbb{P}^n$ by a polynomial q. If A is a point, its polar hyperplane $A^\perp$ with respect to Q is the linear span of the points on Q such that the tangent hyperplanes to Q at these points pass through A. This gives the hyperplane $\sum \frac{\partial q}{\partial X_i}(A)X_i = 0$.

If H is a hyperplane, then its polar point, $H^\perp$, is the intersection of the tangent hyperplanes for Q at the points on $Q \cap H$. Finally, if L is a linear space of dimension d, its polar space $L^\perp$ is the intersection of the polar hyperplanes to points in L. Equivalently, $L^\perp$ is the linear span of all points $H^\perp$, where H is a hyperplane containing L. The dimension of $L^\perp$ is $n - d - 1$.

As an example, consider the case $n = 3$. If A is a point in $\mathbb{P}^3$ then $A^\perp$ is the plane in $\mathbb{P}^3$ defined by the polynomial $\frac{\partial q}{\partial X_0}(A)X_0 + \ldots + \frac{\partial q}{\partial X_3}(A)X_3$. If L is a hyperplane defined by the polynomial $b_0X_0 + b_1X_1 + b_2X_2 + b_3X_3$, then $L^\perp$ is the point A such that

$$(b_0 : b_1 : b_2 : b_3) = \left(\tfrac{\partial q}{\partial X_0}(A) : \tfrac{\partial q}{\partial X_1}(A) : \tfrac{\partial q}{\partial X_2}(A) : \tfrac{\partial q}{\partial X_3}(A)\right)$$

Finally, if L is the line spanned by the points A and B, then $L^\perp$ is the intersection of the two hyperplanes $A^\perp$ and $B^\perp$, i. e.,

$$L^\perp = \mathcal{V}\left(\textstyle\sum_{i=0}^{3} \frac{\partial q}{\partial X_i}(A)X_i\right) \cap \mathcal{V}\left(\textstyle\sum_{i=0}^{3} \frac{\partial q}{\partial X_i}(B)X_i\right)$$

Note that if Q is defined by the polynomial $q = \sum_{i=0}^{n} X_i^2$, then the polar variety of a point $(a_0 : \ldots : a_n) \in \mathbb{P}^n$ is the hyperplane $H = \mathcal{V}(\sum_{i=0}^{n} a_iX_i) \subset \mathbb{P}^n$.

Let $\mathcal{L} : L_0 \subset L_1 \subset \cdots \subset L_{n-1}$ be a flag in $\mathbb{P}^n$, where L_{n-1} is the hyperplane H at infinity if we consider the affine space. We then get the polar flag with respect to Q:

$$\mathcal{L}^\perp : L_{n-1}^\perp \subset L_{n-2}^\perp \subset \cdots \subset L_1^\perp \subset L_0^\perp$$

DEFINITION 2.1 (cf. [3, p. 527]). *The i-th reciprocal polar variety* $W^{\perp}_{L_{i+p-1}}(V)$, $1 \leq i \leq n-p$, *of a variety* V *with respect to the flag* $\mathcal{L}$, *is defined to be the Zariski closure of the set*

$$\{P \in V_{\rm ns} \setminus L^{\perp}_{i+p-1} \mid T_PV \not\pitchfork \langle P, L^{\perp}_{i+p-1}\rangle^{\perp}\}$$

When V is a hypersurface in $\mathbb{P}^n$, the reciprocal polar variety $W^{\perp}_{L_{n-1}}(V)$ is the set $\mathrm{Cl}\{P \in V_{\rm ns} \setminus L^{\perp}_{n-1} \mid T_PV \supset \langle P, L^{\perp}_{n-1}\rangle^{\perp}\}$. In this case, $\langle P, L^{\perp}_{n-1}\rangle$ is the line spanned by P and the point $L^{\perp}_{n-1}$. Since $A^{\perp\perp} = A$, and $A \subseteq B$ implies $A^{\perp} \supseteq B^{\perp}$, it follows that

$$T_PV \supseteq \langle P, L^{\perp}_{n-1}\rangle^{\perp} \Leftrightarrow T_PV^{\perp} \in \langle P, L^{\perp}_{n-1}\rangle.$$

The point $T_PV^{\perp}$ is on the line $\langle P, L^{\perp}_{n-1}\rangle$ if and only if the point $L^{\perp}_{n-1}$ is on the line $\langle P, T_PV^{\perp}\rangle$. So, when V is a hypersurface, the $(n-1)$-th reciprocal variety is

$$W^{\perp}_{L_{n-1}}(V) = \mathrm{Cl}\{P \in V_{\rm ns} \setminus (\{L^{\perp}_{n-1}\} \cup H) \mid L^{\perp}_{n-1} \in \langle P, T_PV^{\perp}\rangle\}.$$

This way of writing the reciprocal polar variety can sometimes be useful, and it gives at better geometric understanding of the reciprocal polar variety.

Let $S \subset \mathbb{A}^n = \mathbb{P}^n \setminus H$ denote the affine part of the variety V, where $H = L_{n-1}$ is the hyperplane at infinity, and let $L \in \mathcal{L}$. We define the *affine* reciprocal polar variety $W^{\perp}_{L}(S)$ to be the affine part $W^{\perp}_{L}(V) \cap \mathbb{A}^n$ of $W^{\perp}_{L}(V)$. The linear variety $\langle P, L^{\perp}\rangle^{\perp}$ is contained in the hyperplane at infinity, so we can consider the affine cone of $\langle P, L^{\perp}\rangle^{\perp}$ as a linear variety $I_{P,L^{\perp}}$ in the affine space. Then the affine reciprocal polar variety can be written as

$$W^{\perp}_{L}(S) = \mathrm{Cl}\{P \in S_{\rm ns} \setminus L^{\perp} \mid t_PS \not\pitchfork I_{P,L^{\perp}}\}$$

where t_PS is the affine tangent space at P, translated to the origin.

The following result, formulated slightly differently, is proved in [3].

PROPOSITION 2.2 ([3, p. 529]). *Let* $S_{\mathbb{R}} \subset \mathbb{A}^n_{\mathbb{R}}$ *be a non-empty, non-singular real variety of pure codimension* p. *Let* $\mathcal{L}$ *be a flag in* $\mathbb{P}^n$, *where* $L_{n-1} = H = \mathbb{P}^n \setminus \mathbb{A}^n$ *is the hyperplane at infinity. Assume* $Q = \mathcal{V}(q)$, *where* q *restricts to a positive definite quadratic form on* $\mathbb{A}^n_{\mathbb{R}}$. *Assume* $H^{\perp} \notin S_{\mathbb{R}}$. *Let* L *be any member of the flag* $\mathcal{L}$ *with* $\dim L \geq p$. *Then the real affine reciprocal polar variety* $W^{\perp}_{L}(S_{\mathbb{R}})$ *contains at least one point from each connected component of* $S_{\mathbb{R}}$.

In Proposition 2.2 we had to assume that $H^{\perp} \notin S_{\mathbb{R}}$ in order to prove that $W^{\perp}_{H}(S_{\mathbb{R}})$ contains a point from each component of $S_{\mathbb{R}}$. The following proposition states that if the variety is a hypersurface and contains $H^{\perp}$, then we can choose another quadric Q', so that the polar point $H^{\perp'}$ with respect to this quadric is not on $S_{\mathbb{R}}$, and we will thus still be able to find points on each component of the hypersurface.

PROPOSITION 2.3. *Let $V \subset \mathbb{P}^n$ be a hypersurface and H a hyperplane, and set $\mathbb{A}^n = \mathbb{P}^n \setminus H$ and $S = V \cap \mathbb{A}^n$. Asssume $S_{\mathbb{R}}$ is non-empty and non-singular. Given any point $A \in \mathbb{A}^n_{\mathbb{R}} \setminus S_{\mathbb{R}}$, there exists a quadric Q' such that the polar point $H^{\perp'}$ with respect to Q' is equal to A, and such that the real affine reciprocal polar variety $W_H^{\perp'}(S_{\mathbb{R}})$ contains at least one point from each connected component of $S_{\mathbb{R}}$.*

Proof. By Proposition 2.2 it suffices to show that we can find $Q' = \mathcal{V}(q')$ such that $H^{\perp'} = A = (1 : a_1 : \ldots : a_n)$ and such that q' restricts to a positive definite form on $\mathbb{A}^n$. Take $q' = (1+\sum_{i=1}^n a_i^2)X_0^2 - 2\sum_{i=1}^n a_i X_0 X_i + \sum_{i=1}^n X_i^2$. Then $\frac{\partial q'}{\partial X_0}(A) = 2$, and $\frac{\partial q'}{\partial X_i}(A) = 0$ for $i = 1, \ldots, n$. The restriction of q' to $\mathbb{A}^n$ is $\sum_{i=1}^n X_i^2$. □

The next proposition provides an explicit description of the affine reciprocal polar variety of a hypersurface $S \subset \mathbb{A}^n$. A similar result is proved in the more general case of a complete intersection variety in [3, 3.1, p. 531].

PROPOSITION 2.4. *Let $V = \mathcal{V}(f) \subset \mathbb{P}^n$ be a real hypersurface, and let $Q = \mathcal{V}(q)$ be defined by an irreducible quadratic polynomial q. We consider the affine space $\mathbb{A}^n = \mathbb{P}^n \setminus H$, where $H = \mathcal{V}(X_0)$. Then the affine reciprocal polar variety $W_H^{\perp}(S)$ of $S = V \cap \mathbb{A}^n$ with respect to Q is equal to the closure of the intersection of $S_{\mathrm{ns}} \setminus \{H^{\perp}\}$ with the variety defined by the 2-minors of the matrix*

$$\begin{pmatrix} \frac{\partial f}{\partial X_1} & \cdots & \frac{\partial f}{\partial X_n} \\ \frac{\partial q}{\partial X_1} & \cdots & \frac{\partial q}{\partial X_n} \end{pmatrix}.$$

Proof. The reciprocal variety $W_H^{\perp}(S)$ is the closure of the set

$$\{P \in S_{\mathrm{ns}} \setminus H^{\perp} \mid I_{P,H^{\perp}} \not\pitchfork t_P S\}.$$

We want to show that for $P = (1 : p_1 : \ldots : p_n) \in S$ the affine cone $I_{P,H^{\perp}}$ over $\langle P, H^{\perp}\rangle^{\perp}$ intersects the tangent space $t_P S$ non-transversally if and only if

$$\operatorname{rank}\begin{pmatrix} \frac{\partial f}{\partial X_1}(P) & \cdots & \frac{\partial f}{\partial X_n}(P) \\ \frac{\partial q}{\partial X_1}(P) & \cdots & \frac{\partial q}{\partial X_n}(P) \end{pmatrix} \leq 1.$$

Note that we have

$$\langle P, H^{\perp}\rangle^{\perp} = P^{\perp} \cap H,$$

so, since $P^{\perp} = \mathcal{V}(\sum_{i=0}^n \frac{\partial q}{\partial X_i}(P)X_i)$ and $H = \mathcal{V}(X_0)$, we find

$$\langle P, H^{\perp}\rangle^{\perp} = \{(0 : X_1 : \ldots : X_n) \mid \sum_{i=1}^n \frac{\partial q}{\partial X_i}(P)X_i = 0\}$$

The affine cone $I_{P,H^{\perp}}$ is the hyperplane $\mathcal{V}(\sum_{i=1}^n \frac{\partial q}{\partial X_i}(P)X_i)$. The affine tangent space $t_P S$ is the hyperplane given by $\sum_{i=1}^n \frac{\partial f}{\partial X_i}(P)X_i = 0$, which implies what we wanted to prove. □

Note that the square of the distance function in the affine space $\mathbb{A}^n$ is given by a quadratic polynomial, so if we let $Q = \mathcal{V}(q)$ be a quadric such that q restricts to this polynomial, we see that the variety defined in Theorem 2 in [7] is nothing but the affine reciprocal polar variety with respect to Q. (This is because elimination of the extra variable gives the 2-minors of the matrix in Proposition 2.4.) Hence [7, Thm. 2] follows from Proposition 2.2.

3. Polar varieties of real singular curves. In this section our varieties will be curves in $\mathbb{P}^2$, so the flags will in this case be of the following form

$$\mathcal{L} : L_0 \subset L_1 \subset \mathbb{P}^2 \text{ and } \mathcal{L}^\perp : L_1^\perp \subset L_0^\perp \subset \mathbb{P}^2$$

where L_0 is a point, and L_1 is the line at infinity, when we look at the affine case. This gives only one interesting classical polar variety and reciprocal polar variety for each choice of flag, namely $W_{L_0}(V)$ and $W_{L_1}^\perp(V)$.

Let $V = \mathcal{V}(f) \subset \mathbb{P}^2$ be a plane curve. The *polar* of V with respect to a point $A = (a_0 : a_1 : a_2)$ is the curve $V' = \mathcal{V}(\sum_{i=0}^2 a_i \frac{\partial f}{\partial X_i})$. It follows from the definition that the polar variety of V with respect to A is equal to

$$W_A(V) = V_{\text{ns}} \cap V'.$$

Given a conic $Q = \mathcal{V}(q)$ and a line L, the *reciprocal polar* of the affine curve $S = V \cap \mathbb{A}^2 \subset \mathbb{A}^2 = \mathbb{P}^2 \setminus L$, is the curve $S'' = \mathcal{V}(\frac{\partial f}{\partial X_1}\frac{\partial q}{\partial X_2} - \frac{\partial f}{\partial X_2}\frac{\partial q}{\partial X_1})$, where f and q are dehomogenized by setting $X_0 = 1$. The affine reciprocal polar variety of S is equal to

$$W_L^\perp(S) = S_{\text{ns}} \cap S''.$$

More generally, we define the *reciprocal polar* of the curve $V \subset \mathbb{P}^2$ with respect to Q and a point $A = (a_0 : a_1 : a_2)$ to be the curve $V'' = \mathcal{V}(\det(f, q, A))$, where $\det(f, q, A)$ is the determinant of the matrix

$$\begin{bmatrix} a_0 & a_1 & a_2 \\ \frac{\partial f}{\partial X_0} & \frac{\partial f}{\partial X_1} & \frac{\partial f}{\partial X_2} \\ \frac{\partial q}{\partial X_0} & \frac{\partial q}{\partial X_1} & \frac{\partial q}{\partial X_2} \end{bmatrix}.$$

The reciprocal polar of the affine curve S is then the affine part of the reciprocal polar with respect to the origin $(1 : 0 : 0)$.

Note that the reciprocal polars form a linear system on $\mathbb{P}^2$ of degree d, whereas the classical polars form a linear system of degree $d-1$. A classical polar variety consists of at most $d(d-1)$ points, whereas a reciprocal polar variety can have as many as d^2 points.

In the following sections we will consider affine curves with singularities, and we determine in which cases the classical polar variety or the reciprocal polar variety will contain non-singular points of each connected component of the real part of the curve.

3.1. Classical polar varieties of real singular affine curves. The classical polar variety of an affine curve $S \subset \mathbb{A}^2$ associated to a given flag $\mathcal{L}$ is the set

$$W_{L_0}(S) = \{P \in S_{\mathrm{ns}} \mid I_1 = t_P S\},$$

where I_1 is the affine line equal to the cone over the point L_0. By an *ordinary real multiple point* we shall mean a singular point with at least two real branches, such that the tangent lines intersect pairwise transversally. If the curve $S_{\mathbb{R}}$ only has ordinary real multiple points as singularities, we have the following proposition.

PROPOSITION 3.1. *Suppose $S_{\mathbb{R}}$ is non-empty and compact and has only ordinary real multiple points as singularities. Then $W_{L_0}(S_{\mathbb{R}})$ contains at least one non-singular point of each connected component.*

Proof. We may assume that $I_1 = \mathcal{V}(X_2)$. Let C be a connected component of $S_{\mathbb{R}}$. The component C has a local maximum point for the coordinate X_2, since C is compact. If this point is a non-singular point, then we know by the proof of Proposition 2.1 that it is contained in the real polar variety. Assume on the contrary that the maximum point P is a singular point, hence an ordinary real multiple point. If $P = (p_1, p_2)$ is a local maximum point for the X_2-coordinate then each of the real branches through P has p_2 as a local maximum for the X_2-coordinate, so the line $\mathcal{V}(X_2 - p_2)$ is a tangent line for each branch. This means that the branches have a common tangent line, hence they do not intersect transversally. □

When it comes to singularities other than ordinary real multiple points, we cannot say whether the singularity can be a maximum point for the X_2-coordinate. But we know that the local maximum and the local minimum points are on the real polar $S'_{\mathbb{R}}$. One point on a component cannot be both a minimum and a maximum unless the component is a line, which is not compact, and therefore excluded. If the singularity is a local maximum point, we know that the minimum point is also on the polar. So we can allow each real component of the curve to have one additional singularity which is not an ordinary real multiple point, and the above result will still remain valid.

For curves with arbitrary singularities, clearly we can have a situation where the points with the maximum and the minimum values for the last coordinate are both singular, with the result that the conclusion of Proposition 3.1 will not be valid for the given choice of flag I_1. One could ask whether it is possible to choose a different affine flag (different affine coordinates) so that the results still holds.

PROPOSITION 3.2. *There exists an affine singular plane curve $S_{\mathbb{R}}$ such that for no choice of flag I_1 does the polar variety contain a non-singular point from each component.*

Proof. We prove this by giving an example of such a curve. Since cusps disturb the continuity of the curvature of a curve, it is natural to look for examples among curves with cusps. So we want to find a curve with at least two components, where each component has cusps. One way to construct such a curve is to consider two, not necessarily irreducible, affine curves $\mathcal{V}(f)$ and $\mathcal{V}(g)$, and then look at the curve $\mathcal{V}(h)$, where $h = f^2 + \epsilon g^3$, which will have cusps at the points of intersection between $\mathcal{V}(f)$ and $\mathcal{V}(g)$. We let $\mathcal{V}(f)$ be the union of two disjoint circles, and $\mathcal{V}(g)$ the union of four lines, where two of the lines intersect one of the circles twice, and the other two lines intersect the other circle twice. We can take

$$f = (X_1^2 + X_2^2 - 1)((X_1 - 4)^2 + (X_2 - 2)^2 - 1)$$

and

$$g = (X_2 - \tfrac{1}{2})(X_2 + \tfrac{1}{2})(X_1 - \tfrac{7}{2})(X_1 - \tfrac{9}{2}).$$

The curve $\mathcal{V}(h) = \mathcal{V}(f^2 + \frac{1}{100}g^3)$ has then four compact components with two cusps on each component.

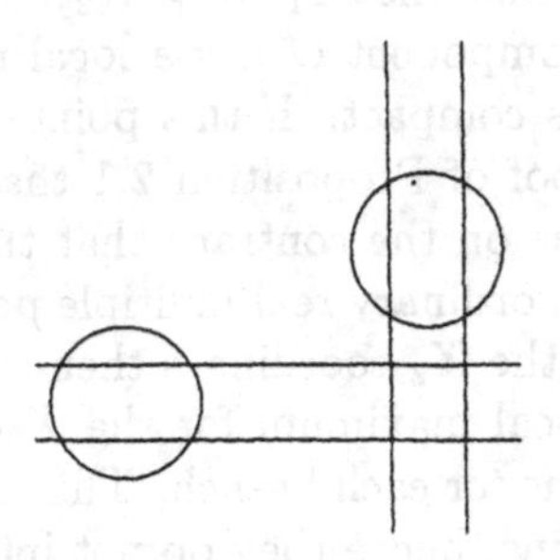

FIG. 2. *The curves $\mathcal{V}(f)$ and $\mathcal{V}(g)$.*

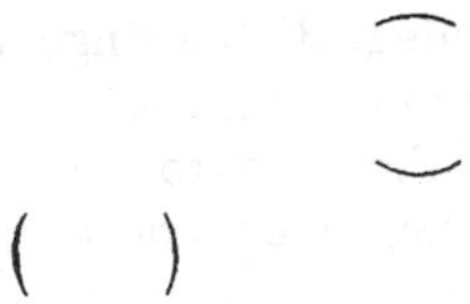

FIG. 3. *The curve $\mathcal{V}(f^2 + \frac{1}{100}g^3)$.*

A non-singular point P is on the affine polar variety with respect to a line $I = \mathcal{V}(aX_1 + bX_2)$ if the tangent line $\mathcal{V}(\frac{\partial f}{\partial X_1}(P)X_1 + \frac{\partial f}{\partial X_2}(P)X_2)$ is equal to I. Consider the Gauss map $\gamma\colon S_{\mathbb{R}} \to \mathbb{P}^1_{\mathbb{R}}$ given by

$$P \mapsto (\tfrac{\partial f}{\partial X_1}(P) : \tfrac{\partial f}{\partial X_2}(P)).$$

Let C_i, $i = 1, \ldots, 4$, denote the connected components of $S_{\mathbb{R}}$. Let us show that $\cap_{i=1}^4 \gamma(C_i) = \emptyset$.

The two lower components have cusps at the points $(-\frac{\sqrt{3}}{2}, \pm\frac{1}{2})$ and $(\frac{\sqrt{3}}{2}, \pm\frac{1}{2})$, and the map γ sends these points to the lines $\mathcal{V}(3X_1 \mp \sqrt{3}X_2)$ and $\mathcal{V}(3X_1 \pm \sqrt{3}X_2)$ respectively. Each of these components have points that are mapped to the line $\mathcal{V}(X_1)$ by γ. The components do not have any inflection points, so the image of the map γ varies continuously from $\mathcal{V}(X_1)$ to $\mathcal{V}(3X_1 + \sqrt{3}X_2)$ and $\mathcal{V}(3X_1 - \sqrt{3}X_2)$, hence the other tangent lines lie in the sector between the two lines $\mathcal{V}(3X_1 + \sqrt{3}X_2)$ and $\mathcal{V}(3X_1 - \sqrt{3}X_2)$.

The same happens if we calculate the tangent lines at the cusps for the two upper components. The tangent lines of these components have to be in the sector between $\mathcal{V}(3X_2 + \sqrt{3}X_1)$ and $\mathcal{V}(3X_2 - \sqrt{3}X_1)$, which contains the line $\mathcal{V}(X_2)$. These two sectors do not intersect except at the origin, so for any choice of line $I = \mathcal{V}(aX_1 + bX_2)$ the set $(\gamma^{-1}(I)) \cap C$ is empty or consists of points of the two lower components or points of the two upper components. □

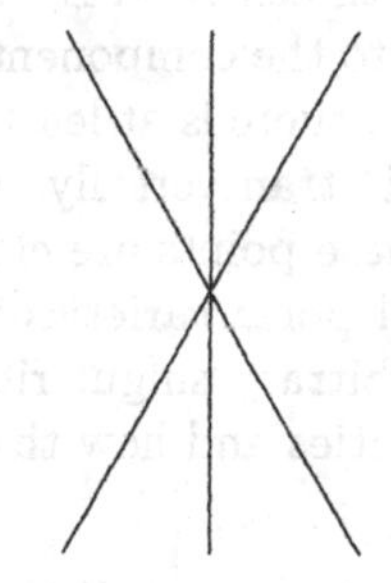

FIG. 4. *The lines* $\mathcal{V}(3X_1 + \sqrt{3}X_2)$, $\mathcal{V}(3X_1 - \sqrt{3}X_2)$, *and* $\mathcal{V}(X_1)$.

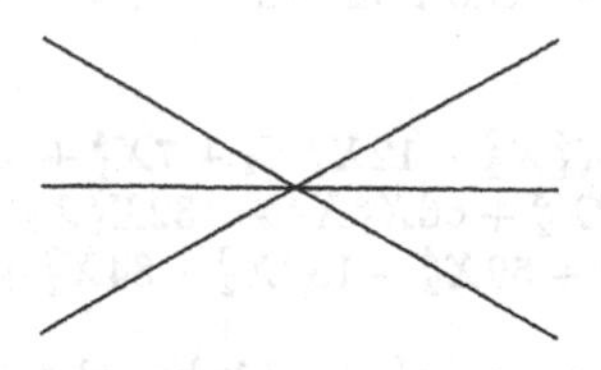

FIG. 5. *The lines* $\mathcal{V}(3X_2 + \sqrt{3}X_1)$, $\mathcal{V}(3X_2 - \sqrt{3}X_1)$, *and* $\mathcal{V}(X_2)$.

3.2. Reciprocal polar varieties of affine real singular curves. As for the classical polar varieties we will here determine whether Proposition 2.2 is valid for affine curves with singularities. We can easily prove the following for curves with ordinary multiple points.

PROPOSITION 3.3. *Let $S \subset \mathbb{A}^2 = \mathbb{P}^2 \setminus L$ be an affine plane curve. Suppose that $S_\mathbb{R}$ is non-empty and has no other singularities than ordinary* *real multiple points. Further, let Q be defined by a polynomial q which restricts to a positive definite quadratic form on $\mathbb{A}^2_{\mathbb{R}}$. Assume that $L^\perp$ is not contained in $S_{\mathbb{R}}$. Then the real affine reciprocal polar variety $W^\perp_L(S_{\mathbb{R}})$ contains at least one non-singular point from each connected component of $S_{\mathbb{R}}$.*

Proof. Note that the hypotheses imply that $L^\perp \in \mathbb{A}^2 = \mathbb{P}^2 \setminus L$. Moreover, the restriction of the quadratic polynomial q to $\mathbb{A}^2$ defines a distance function on $\mathbb{A}^2_{\mathbb{R}}$. The proof of Proposition 2.2 given in [3] consists, in this case, of showing that for each component, a point with the shortest distance to the point $L^\perp$ is a point on the reciprocal polar variety. So, we must show that if the component contains ordinary real multiple points, then these points can not be among the points with the shortest distance to the point $L^\perp$.

Assume on the contrary that, for a given component C of the curve $S_{\mathbb{R}}$, there is an ordinary real multiple point P which has the shortest distance to the point $L^\perp$. The conic with centre in $L^\perp$ and radius $\mathrm{dist}(L^\perp, P)$ will by our assumption be tangent to the component at the point P. Since P is an ordinary real multiple point, there is at least one other real branch, and this branch intersects the conic transversally. Hence this branch contains points inside the conic, and these points are closer to $L^\perp$ than P. □

As in the case of classical polar varieties we cannot prove the above proposition for curves with arbitrary singularities, since the situation will depend on the type of singularities and how the singularities are placed on the component.

4. Examples. In this section we shall look at some examples of singular plane affine curves and their polars and reciprocal polars, thus illustrating the propositions in the previous sections. We use SURF [5] to draw the curves.

EXAMPLE 1. Consider the real affine curve S_1 of degree 6 defined by the polynomial

$$\begin{aligned} f_1 := \; & X_1^6 + 3X_1^4X_2^2 - 12X_1^4X_2 + 7X_1^4 + 3X_1^2X_2^4 \\ & - 24X_1^2X_2^3 + 66X_1^2X_2^2 - 132X_1^2X_2 + 136X_1^2 + X_2^6 \\ & - 12X_2^5 + 59X_2^4 - 132X_2^3 + 84X_2^2 + 144X_2 - 143. \end{aligned}$$

This curve is compact and smooth, and it has three connected components.

We let $\mathcal{L}$ be the flag

$$L_0 = \mathcal{V}(X_0, X_1) \subset L_1 = \mathcal{V}(X_0).$$

The real polar variety is the set

$$W_{L_0}((S_1)_{\mathbb{R}}) = \{P \in (S_1)_{\mathbb{R}} \mid \tfrac{\partial f_1}{\partial X_2}(P) = 0\},$$

which is equal to intersection of $(S_1)_{\mathbb{R}}$ and its polar $(S_1')_{\mathbb{R}} = \mathcal{V}(\frac{\partial f_1}{\partial X_2})_{\mathbb{R}}$. We see that the polar variety contains points from each connected component

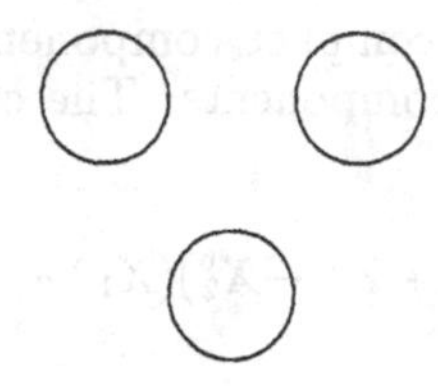

FIG. 6. *The curve S_1.*

and that the points of the affine polar variety are exactly those points on each component which give maximal and minimal values for the X_1-coordinate.

FIG. 7. *The curve S_1 and its polar S_1'.*

Let Q be the standard quadric, given by $q = \sum_{i=0}^{2} X_i = 0$. The real affine reciprocal polar variety consists of the real points of the intersection between S_1 and its reciprocal polar $S_1'' = \mathcal{V}(X_2 \frac{\partial f_1}{\partial X_1} - X_1 \frac{\partial f_1}{\partial X_2})$, and we know from Proposition 2.2 that also the real affine reciprocal polar variety contains points from each connected component. The reciprocal polar variety consists of the points on each component with the locally shortest or longest Euclidean distance to the origin.

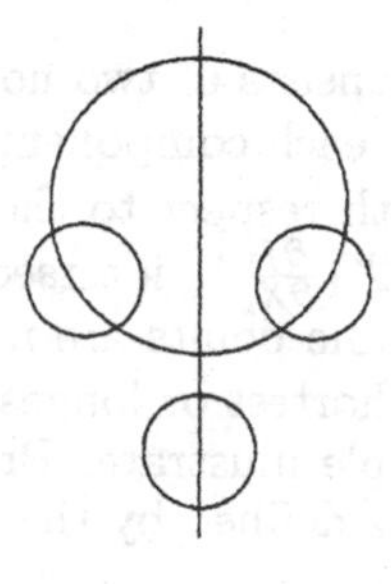

FIG. 8. *The curve S_1 and its reciprocal polar S_1''.*

EXAMPLE 2. To illustrate Proposition 3.1 we consider an affine irreducible curve S_2 with two compact components and with one ordinary double point on each of the components. The curve is given by the polynomial

$$f_2 = ((X_1+2)X_2 - (X_1+2)^6 - X_2^6)(X_1X_2 - X_1^6 - X_2^6) + \tfrac{1}{100}X_2^6.$$

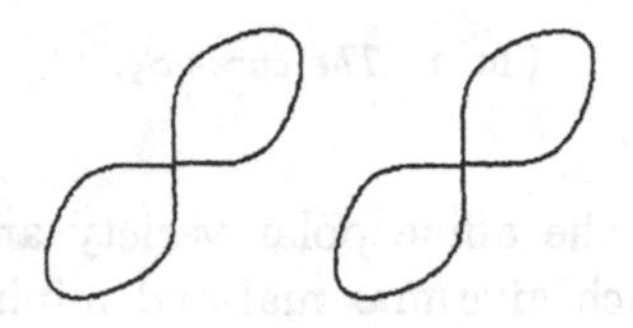

FIG. 9. *The curve S_2.*

We let $\mathcal{L}$ be as in the example above, and we see that the affine polar variety contains non-singular points from each of the components of the curve.

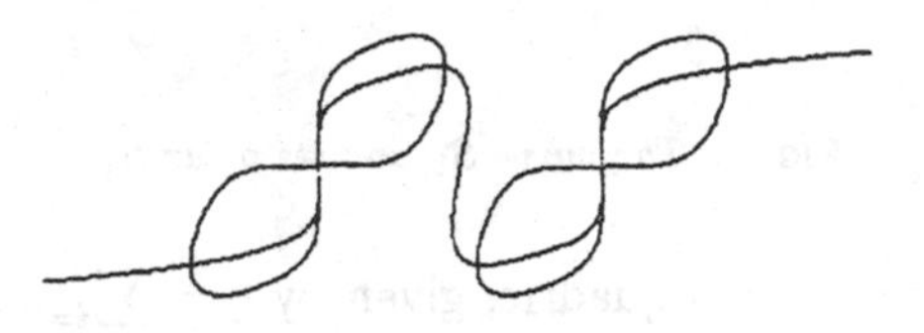

FIG. 10. *The curve S_2 and its polar S_2'.*

EXAMPLE 3. Consider the irreducible curve $S_3 = \mathcal{V}(f_3)$, where

$$f_3 = \mathcal{V}(144 - 24X_2^2 - 88X_1^2 + X_2^4 - X_1^6 + 17X_1^4 - 14X_2^2X_1^2 + \tfrac{1}{100}X_2^6).$$

The real part of the curve consists of two non-compact components with one ordinary double point on each component.

The reciprocal polar, with respect to the standard flag and standard quadric, $S_3'' = \mathcal{V}(X_2\frac{\partial f_3}{\partial X_1} - X_1\frac{\partial f_3}{\partial X_2})$, intersects each component in non-singular points, since the double points are not among the points on each component with the locally shortest or longest distance from the origin.

EXAMPLE 4. This example illustrates Proposition 2.3. We start with the irreducible affine curve S_4 defined by the polynomial

$$f_4 := X_1^2 - X_2(X_2+1)(X_2+2).$$

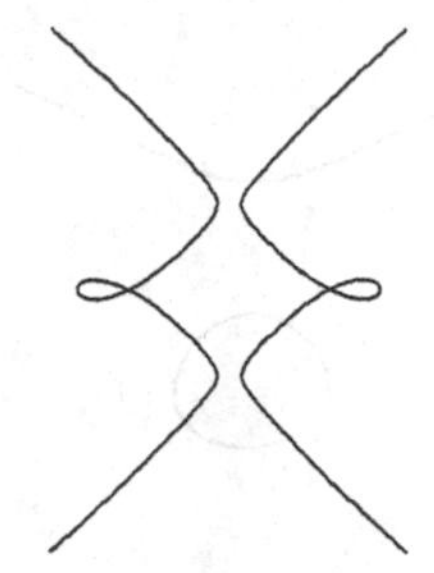

FIG. 11. *The curve S_3.*

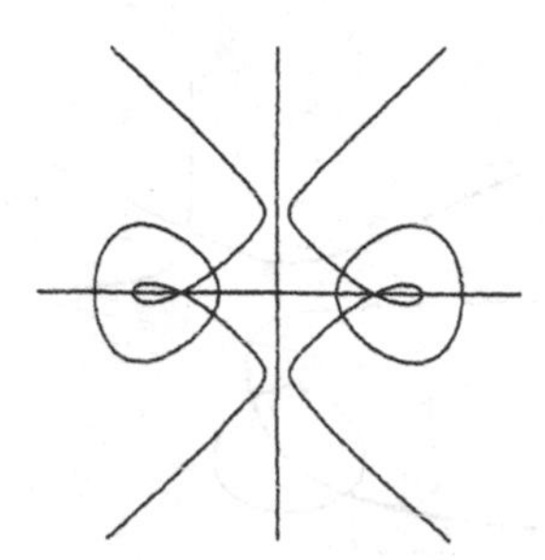

FIG. 12. *The curve S_3 and its reciprocal polar S_3''.*

This curve passes through the origin, so the reciprocal polar variety, when Q is the standard quadric and $L = \mathcal{V}(X_0)$ is the line at infinity, will not contain any points from the connected component containing the origin, since we are not counting points which are on both the variety and the flag. Instead we choose the point $(1,0)$ (or $(1:1:0)$ in projective coordinates), and we must find a polynomial q' such that $(1:1:0)^{\perp'}$ is the line $\mathcal{V}(X_0)$. We see that the polynomial $2X_0^2 - 2X_0X_1 + X_1^2 + X_2^2$ will do, and we will now find the reciprocal polar variety $W_L^{\perp'}(S_4)$. If $P = (1:p_1:p_2)$, the point $\langle P, L^{\perp'}\rangle^{\perp'} = P^{\perp'} \cap L$ is the point $(0:p_2:1-p_1)$, so the affine cone over it, $I_{P,L^{\perp'}}$, is the line $\mathcal{V}((p_1-1)X_1 + p_2X_2)$. The reciprocal polar variety $W_L^{\perp'}(S_4)$ is the set

$$\{P \in (S_4)_{\mathbb{R}} \mid p_2 \tfrac{\partial f_4}{\partial X_1}(P) + (1-p_1)\tfrac{\partial f_4}{\partial X_2}(P) = 0\};$$

this set consists of the points on $(S_4)_{\mathbb{R}}$ with the locally shortest or longest Euclidean distance to the point $(1,0)$, and it contains points from each component of $(S_4)_{\mathbb{R}}$.

EXAMPLE 5. This is an example showing that Proposition 3.3 does not hold for curves with arbitrary singularities. Consider the curve S_5 given by the polynomial

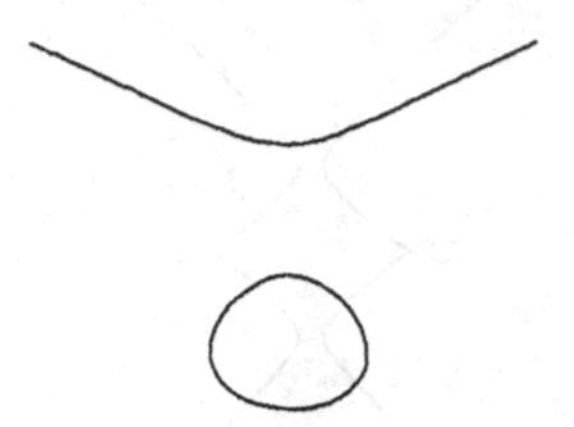

FIG. 13. *The curve S_4.*

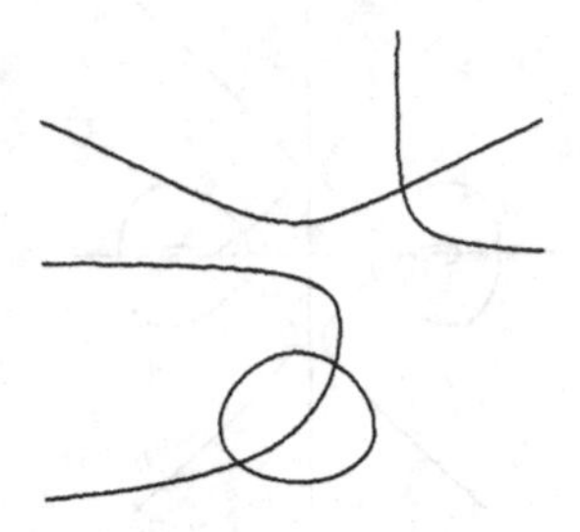

FIG. 14. *The curve S_4 and its reciprocal polar S_4''.*

$$f_5 = ((X_1 - 4)^2 + (X_2 - 2)^2 - 1)^2 + \tfrac{1}{100}((X_1 - \tfrac{7}{2})(X_1 - \tfrac{9}{2}))^3,$$

The real components of this curve do not contain points on the reciprocal polar variety other than the four cusps. This can be seen by calculating the intersection points between S_5 and its reciprocal polar $S_5'' = \mathcal{V}(X_1 \frac{\partial f_5}{\partial X_2} - X_2 \frac{\partial f_5}{\partial X_1})$.

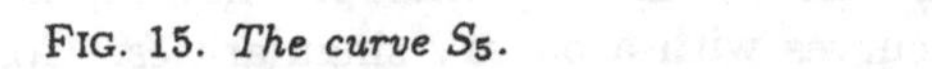

FIG. 15. *The curve S_5.*

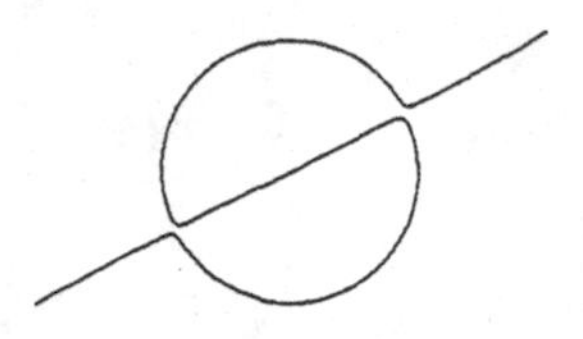

FIG. 16. *The reciprocal polar* S_5''.

REFERENCES

[1] B. BANK, M. GIUSTI, J. HEINTZ, AND G.M. MBAKOP. Polar varieties, real equation solving, and data structures: The hypersurface case. *J. Complexity*, **13**(1): 5–27, 1997.

[2] B. BANK, M. GIUSTI, J. HEINTZ, AND G.M. MBAKOP. Polar varieties and efficient real elimination. *Math. Z.*, **238**(1): 115–144, 2001.

[3] B. BANK, M. GIUSTI, J. HEINTZ, AND L.M. PARDO. Generalized polar varieties and an efficient real elimination procedure. *Kybernetika (Prague)*, **40**(5): 519–550, 2004.

[4] B. BANK, M. GIUSTI, J. HEINTZ, AND L.M. PARDO. Generalized polar varieties: Geometry and algorithms. *J. Complexity*, **21**(4): 377–412, 2005.

[5] S. ENDRASS et al. Surf 1.0.4. 2003. A Computer Software for Visualising Real Algebraic Geometry, `http://surf.sourceforge.net`.

[6] R. PIENE. Polar classes of singular varieties. *Ann. Sci. École Norm. Sup. (4)*, **11**(2): 247–276, 1978.

[7] M. SAFEY EL DIN. Putting into practice deformation techniques for the computation of sampling points in real singular hypersurfaces. 2005.

[8] M. SAFEY EL DIN AND É. SCHOST. Polar varieties and computation of one point in each connected component of a smooth algebraic set. In *Proceedings of the 2003 International Symposium on Symbolic and Algebraic Computation*, pp. 224–231 (electronic), New York, 2003. ACM.

SEMIDEFINITE REPRESENTATION OF THE K-ELLIPSE

JIAWANG NIE*, PABLO A. PARRILO†, AND BERND STURMFELS‡

Abstract. The k-ellipse is the plane algebraic curve consisting of all points whose sum of distances from k given points is a fixed number. The polynomial equation defining the k-ellipse has degree 2^k if k is odd and degree $2^k - \binom{k}{k/2}$ if k is even. We express this polynomial equation as the determinant of a symmetric matrix of linear polynomials. Our representation extends to weighted k-ellipses and k-ellipsoids in arbitrary dimensions, and it leads to new geometric applications of semidefinite programming.

Key words. k-ellipse, algebraic degree, semidefinite representation, Zariski closure, tensor sum.

AMS(MOS) subject classifications. 90C22, 52A20, 14Q10.

1. Introduction.

The *circle* is the plane curve consisting of all points (x, y) whose distance from a given point (u_1, v_1) is a fixed number d. It is the zero set of the quadratic polynomial

$$p_1(x,y) \quad = \quad \det \begin{bmatrix} d+x-u_1 & y-v_1 \\ y-v_1 & d-x+u_1 \end{bmatrix}. \tag{1.1}$$

The *ellipse* is the plane curve consisting of all points (x, y) whose sum of distances from two given points (u_1, v_1) and (u_2, v_2) is a fixed number d. It is the zero set of

$$p_2(x,y) = \det \begin{bmatrix} d+2x-u_1-u_2 & y-v_1 & y-v_2 & 0 \\ y-v_1 & d+u_1-u_2 & 0 & y-v_2 \\ y-v_2 & 0 & d-u_1+u_2 & y-v_1 \\ 0 & y-v_2 & y-v_1 & d-2x+u_1+u_2 \end{bmatrix}. \tag{1.2}$$

In this paper we generalize these determinantal formulas for the circle and the ellipse. Fix a positive real number d and fix k distinct points $(u_1, v_1), (u_2, v_2), \ldots, (u_k, v_k)$ in $\mathbb{R}^2$. The *k-ellipse* with *foci* (u_i, v_i) and *radius* d is the following curve in the plane:

$$\left\{ (x,y) \in \mathbb{R}^2 \, : \, \sum_{i=1}^{k} \sqrt{(x-u_i)^2 + (y-v_i)^2} \, = \, d \right\}. \tag{1.3}$$

The k-ellipse is the boundary of a convex set $\mathcal{E}_k$ in the plane, namely, the set of points whose sum of distances to the k given points is at most d.

*Department of Mathematics, University of California at San Diego, La Jolla, CA 92093.

†Laboratory for Information and Decision Systems, Massachusetts Institute of Technology, Cambridge, MA 02139.

‡Department of Mathematics, University of California at Berkeley, Berkeley, CA 94720.

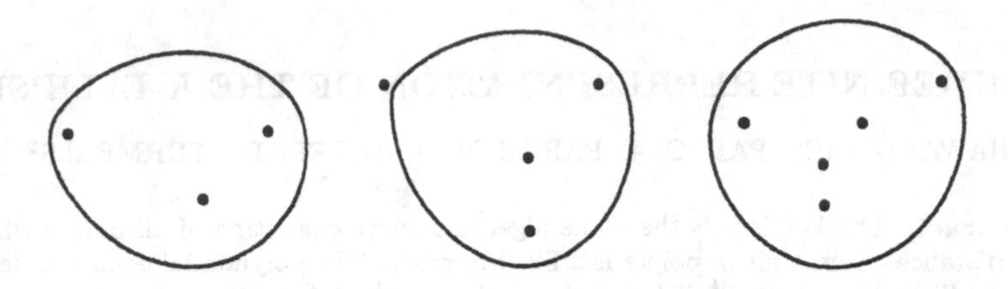

FIG. 1. *A 3-ellipse, a 4-ellipse, and a 5-ellipse, each with its foci.*

These convex sets are of interest in computational geometry [16] and in optimization, e.g. for the Fermat-Weber facility location problem [1, 3, 7, 14, 18]. In the classical literature (e.g. [15]), k-ellipses are known as *Tschirnhaus'sche Eikurven* [11]. Indeed, they look like "egg curves" and they were introduced by Tschirnhaus in 1686.

We are interested in the irreducible polynomial $p_k(x, y)$ that vanishes on the k-ellipse. This is the unique (up to sign) polynomial with coprime integer coefficients in the unknowns x and y and the parameters $d, u_1, v_1, \ldots, u_k, v_k$. By the *degree of the k-ellipse* we mean the degree of $p_k(x, y)$ in x and y. To compute it, we must eliminate the square roots in the representation (1.3). Our solution to this problem is as follows:

THEOREM 1.1. *The k-ellipse has degree 2^k if k is odd and degree $2^k - \binom{k}{k/2}$ if k is even. Its defining polynomial has a determinantal representation*

$$p_k(x,y) \quad = \quad \det\left(x \cdot A_k + y \cdot B_k + C_k \right) \tag{1.4}$$

where A_k, B_k, C_k are symmetric $2^k \times 2^k$ matrices. The entries of A_k and B_k are integer numbers, and the entries of C_k are linear forms in the parameters $d, u_1, v_1, \ldots, u_k, v_k$.

For the circle ($k = 1$) and the ellipse ($k = 2$), the representation (1.4) is given by the formulas (1.1) and (1.2). The polynomial $p_3(x, y)$ for the 3-ellipse is the determinant of

$$\begin{bmatrix}
d{+}3x{-}u_1{-}u_2{-}u_3 & y{-}v_1 & y{-}v_2 & 0 \\
y{-}v_1 & d{+}x{+}u_1{-}u_2{-}u_3 & 0 & y{-}v_2 \\
y{-}v_2 & 0 & d{+}x{-}u_1{+}u_2{-}u_3 & y{-}v_1 \\
0 & y{-}v_2 & y{-}v_1 & d{-}x{+}u_1{+}u_2{-}u_3 \\
y{-}v_3 & 0 & 0 & 0 \\
0 & y{-}v_3 & 0 & 0 \\
0 & 0 & y{-}v_3 & 0 \\
0 & 0 & 0 & y{-}v_3
\end{bmatrix}$$

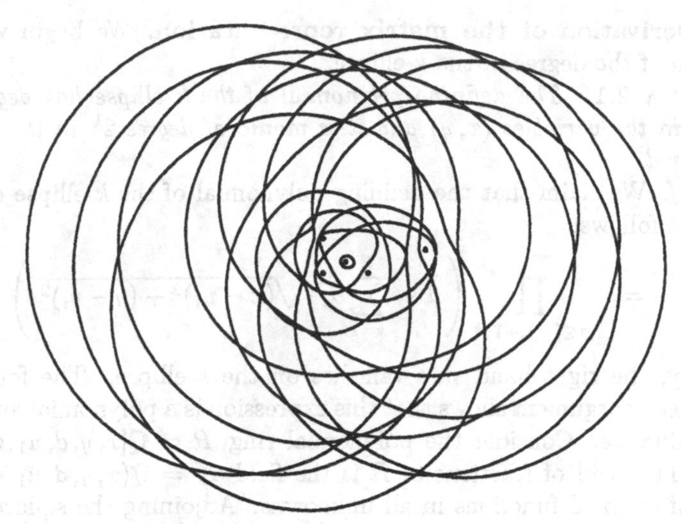

FIG. 2. *The Zariski closure of the 5-ellipse is an algebraic curve of degree 32. The tiny curve in the center is the 5-ellipse.*

$$
\left.\begin{array}{cccc}
y-v_3 & 0 & 0 & 0 \\
0 & y-v_3 & 0 & 0 \\
0 & 0 & y-v_3 & 0 \\
0 & 0 & 0 & y-v_3 \\
d+x-u_1-u_2+u_3 & y-v_1 & y-v_2 & 0 \\
y-v_1 & d-x+u_1-u_2+u_3 & 0 & y-v_2 \\
y-v_2 & 0 & d-x-u_1+u_2+u_3 & y-v_1 \\
0 & y-v_2 & y-v_1 & d-3x+u_1+u_2+u_3
\end{array}\right] .
$$

The full expansion of this 8×8-determinant has 2,355 terms. Next, the 4-ellipse is a curve of degree ten which is represented by a symmetric 16×16-matrix, etc....

This paper is organized as follows. The proof of Theorem 1.1 will be given in Section 2. Section 3 is devoted to geometric aspects and connections to semidefinite programming. While the k-ellipse itself is a curve of convex shape, its Zariski closure $\{\, p_k(x,y) = 0 \,\}$ has many extra branches outside the convex set $\mathcal{E}_k$. They are arranged in a beautiful pattern known as a *Helton-Vinnikov curve* [5]. This pattern is shown in Figure 2 for $k=5$ points. In Section 4 we generalize our results to the weighted case and to higher dimensions, and we discuss the computation of the *Fermat-Weber point* of the given points (u_i, v_i). A list of open problems and future directions is presented in Section 5.

2. Derivation of the matrix representation. We begin with a discussion of the degree of the k-ellipse.

LEMMA 2.1. *The defining polynomial of the k-ellipse has degree at most 2^k in the variables (x, y) and it is monic of degree 2^k in the radius parameter d.*

Proof. We claim that the defining polynomial of the k-ellipse can be written as follows:

$$p_k(x,y) \quad = \quad \prod_{\sigma \in \{-1,+1\}^k} \left(d - \sum_{i=1}^{k} \sigma_i \cdot \sqrt{(x-u_i)^2 + (y-v_i)^2} \right). \quad (2.1)$$

Obviously, the right hand side vanishes on the k-ellipse. The following Galois theory argument shows that this expression is a polynomial and that it is irreducible. Consider the polynomial ring $R = \mathbb{Q}[x, y, d, u_1, v_1, \ldots, u_k, v_k]$. The field of fractions of R is the field $K = \mathbb{Q}(x, y, d, u_1, v_1, \ldots, u_k, v_k)$ of rational functions in all unknowns. Adjoining the square roots in (1.3) to K gives an algebraic field extension L of degree 2^k over K. The Galois group of the extension L/K is $(\mathbb{Z}/2\mathbb{Z})^k$, and the product in (2.1) is over the orbit of the element $d - \sum_{i=1}^{k} \sqrt{(x-u_i)^2 + (y-v_i)^2}$ of L under the action of the Galois group. Thus this product in (2.1) lies in the ground field K. Moreover, each factor in the product is integral over R, and therefore the product lies in the polynomial ring R. To see that this polynomial is irreducible, it suffices to observe that no proper subproduct of the right hand side in (2.1) lies in the ground field K. □

The statement *degree at most* 2^k is the crux in Lemma 2.1. Indeed, the degree in (x, y) can be strictly smaller than 2^k as the case of the classical ellipse ($k = 2$) demonstrates. When evaluating the product (2.1) some unexpected cancellations may occur. This phenomenon happens for all even k, as we shall see later in this section.

We now turn to the matrix representation promised by Theorem 1.1. We recall the following standard definition from matrix theory (e.g., [6]). Let A be a real $m \times m$-matrix and B a real $n \times n$-matrix. The *tensor sum* of A and B is the $mn \times mn$ matrix $A \oplus B := A \otimes I_n + I_m \otimes B$. The tensor sum of square matrices is an associative operation which is not commutative. For instance, for three matrices A, B, C we have

$$A \oplus B \oplus C \quad = \quad A \otimes I \otimes I + I \otimes B \otimes I + I \otimes I \otimes C.$$

Here $\otimes$ denotes the *tensor product*, which is also associative but not commutative. Tensor products and tensor sums of matrices are also known as *Kronecker products* and *Kronecker sums* [2, 6]. Tensor sums of symmetric matrices can be diagonalized by treating the summands separately:

LEMMA 2.2. *Let $M_1, \ldots, M_k$ be symmetric matrices, let $U_1, \ldots, U_k$ be orthogonal matrices, and let $\Lambda_1, \ldots, \Lambda_k$ be diagonal matrices such that $M_i = U_i \cdot \Lambda_i \cdot U_i^T$ for $i = 1, \ldots, k$. Then*

$$(U_1 \otimes \cdots \otimes U_k)^T \cdot (M_1 \oplus \cdots \oplus M_k) \cdot (U_1 \otimes \cdots \otimes U_k) = \Lambda_1 \oplus \cdots \oplus \Lambda_k.$$

In particular, the eigenvalues of the tensor sum $M_1 \oplus M_2 \oplus \cdots \oplus M_k$ are the sums $\lambda_1 + \lambda_2 + \cdots + \lambda_k$ where λ_1 is any eigenvalue of M_1, λ_2 is any eigenvalue of M_2, etc.

The proof of this lemma is an exercise in (multi)-linear algebra. We are now prepared to state our formula for the explicit determinantal representation of the k-ellipse.

THEOREM 2.1. *Given points $(u_1, v_1), \ldots, (u_k, v_k)$ in $\mathbb{R}^2$, we define the $2^k \times 2^k$ matrix*

$$L_k(x,y) := d \cdot I_{2^k} + \begin{bmatrix} x-u_1 & y-v_1 \\ y-v_1 & -x+u_1 \end{bmatrix} \oplus \cdots \oplus \begin{bmatrix} x-u_k & y-v_k \\ y-v_k & -x+u_k \end{bmatrix} \quad (2.2)$$

which is affine in x, y and d. Then the k-ellipse has the determinantal representation

$$p_k(x,y) \;=\; \det L_k(x,y). \quad (2.3)$$

The convex region bounded by the k-ellipse is defined by the following matrix inequality:

$$\mathcal{E}_k \;=\; \{\, (x,y) \in \mathbb{R}^2 \;:\; L_k(x,y) \succeq 0 \,\}. \quad (2.4)$$

Proof. Consider the 2×2 matrix that appears as a tensor summand in (2.2):

$$\begin{bmatrix} x-u_i & y-v_i \\ y-v_i & -x+u_i \end{bmatrix}.$$

A computation shows that this matrix is orthogonally similar to

$$\begin{bmatrix} \sqrt{(x-u_i)^2 + (y-v_i)^2} & 0 \\ 0 & -\sqrt{(x-u_i)^2 + (y-v_i)^2} \end{bmatrix}.$$

These computations take place in the field L which was considered in the proof of Lemma 2.1 above. Lemma 2.2 is valid over any field, and it implies that the matrix $L_k(x,y)$ is orthogonally similar to a $2^k \times 2^k$ diagonal matrix with diagonal entries

$$d + \sum_{i=1}^{k} \sigma_i \cdot \sqrt{(x-u_i)^2 + (y-v_i)^2}, \qquad \sigma_i \in \{-1, +1\}. \quad (2.5)$$

The desired identity (2.3) now follows directly from (2.1) and the fact that the determinant of a matrix is the product of its eigenvalues. For the characterization of the convex set $\mathcal{E}_k$, notice that positive semidefiniteness of $L_k(x,y)$ is equivalent to nonnegativity of all the eigenvalues (2.5). It suffices to consider the smallest eigenvalue, which equals

$$d - \sum_{i=1}^{k} \sqrt{(x-u_i)^2 + (y-v_i)^2}.$$

Indeed, this quantity is nonnegative if and only if the point (x, y) lies in $\mathcal{E}_k$. □

Proof. [Proof of Theorem 1.1] The assertions in the second and third sentence have just been proved in Theorem 2.1. What remains to be shown is the first assertion concerning the degree of $p_k(x,y)$ as a polynomial in (x,y). To this end, we consider the univariate polynomial $g(t) := p_k(t\cos\theta, t\sin\theta)$ where θ is a generic angle. We must prove that

$$\deg_t(g(t)) \;=\; \begin{cases} 2^k & \text{if } k \text{ is odd,} \\ 2^k - \binom{k}{k/2} & \text{if } k \text{ is even.} \end{cases}$$

The polynomial $g(t)$ is the determinant of the symmetric $2^k \times 2^k$-matrix

$$L_k(t\cos\theta, t\sin\theta) = t\cdot\left(\begin{bmatrix} \cos\theta & \sin\theta \\ \sin\theta & -\cos\theta \end{bmatrix} \oplus \cdots \oplus \begin{bmatrix} \cos\theta & \sin\theta \\ \sin\theta & -\cos\theta \end{bmatrix}\right) + C_k. \quad (2.6)$$

The matrix C_k does not depend on t. We now define the $2^k \times 2^k$ orthogonal matrix

$$U \;:=\; \underbrace{V \otimes \cdots \otimes V}_{k \text{ times}} \qquad \text{where} \quad V := \begin{bmatrix} \cos(\theta/2) & -\sin(\theta/2) \\ \sin(\theta/2) & \cos(\theta/2) \end{bmatrix},$$

and we note the matrix identity

$$V^T \cdot \begin{bmatrix} \cos\theta & \sin\theta \\ \sin\theta & -\cos\theta \end{bmatrix} \cdot V \;=\; \begin{bmatrix} 1 & 0 \\ 0 & -1 \end{bmatrix}.$$

Pre- and post-multiplying (2.6) by U^T and U, we find that

$$U^T \cdot L_k(t\cos\theta, t\sin\theta) \cdot U = t \cdot \underbrace{\left(\begin{bmatrix} 1 & 0 \\ 0 & -1 \end{bmatrix} \oplus \cdots \oplus \begin{bmatrix} 1 & 0 \\ 0 & -1 \end{bmatrix}\right)}_{E_k} + U^T \cdot C_k \cdot U.$$

The determinant of this matrix is our univariate polynomial $g(t)$. The matrix E_k is a diagonal matrix of dimension $2^k \times 2^k$. Its diagonal entries are obtained by summing k copies of -1 or $+1$ in all 2^k possible ways. None of these sums are zero when k is odd, and precisely $\binom{k}{k/2}$ of these sums are zero when k is even. This shows that the rank of E_k is 2^k when k is odd, and it is $2^k - \binom{k}{k/2}$ when k is even. We conclude that the univariate polynomial $g(t) = \det\big(t \cdot E_k + U^T C_k U\big)$ has the desired degree. □

3. More pictures and some semidefinite aspects. In this section we examine the geometry of the k-ellipse, we look at some pictures, and we discuss aspects relevant to the theory of semidefinite programming. In Figure 1 several k-ellipses are shown, for $k = 3, 4, 5$. One immediately observes that, in contrast to the classical circle and ellipse, a k-ellipse does

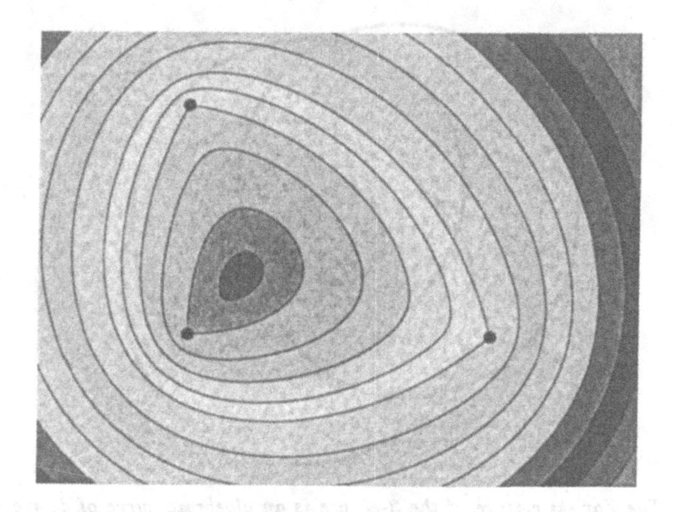

FIG. 3. *A pencil of 3-ellipses with fixed foci (the three bold dots) and different radii. These 3-elliptical curves are always smooth unless they contain one of the foci.*

not necessarily contain the foci in its interior. The interior $\mathcal{E}_k$ of the k-ellipse is a sublevel set of the convex function

$$(x,y) \quad \mapsto \quad \sum_{i=1}^{k} \sqrt{(x-u_i)^2 + (y-v_i)^2}. \tag{3.1}$$

This function is strictly convex, provided the foci $\{(u_i, v_i)\}_{i=1}^{k}$ are not collinear [14]. This explains why the k-ellipse is a convex curve. In order for $\mathcal{E}_k$ to be nonempty, it is necessary and sufficient that the radius d be greater than or equal to the global minimum d_* of the convex function (3.1).

The point (x_*, y_*) at which the global minimum d_* is achieved is called the *Fermat-Weber point* of the foci. This point minimizes the sum of the distances to the k given points (u_i, v_i), and it is of importance in the facility location problem. See [1, 3, 7], and [15] for a historical reference. For a given set of foci, we can vary the radius d, and this results in a pencil of confocal k-ellipses, as in Figure 3. The sum of distances function (3.1) is differentiable everywhere except at the (u_i, v_i), where the square root function has a singularity. As a consequence, the k-ellipse is a smooth curve of convex shape, except when that curve contains one of the foci.

An algebraic geometer would argue that there is more to the k-ellipse than meets the eye in Figures 1 and 3. We define the *algebraic k-ellipse* to be the Zariski closure of the k-ellipse, or, equivalently, the zero set of the polynomial $p_k(x,y)$. The algebraic k-ellipse is an algebraic curve, and it can be considered in either the real plane $\mathbb{R}^2$, in the complex plane $\mathbb{C}^2$, or (even better) in the complex projective plane $\mathbb{CP}^2$.

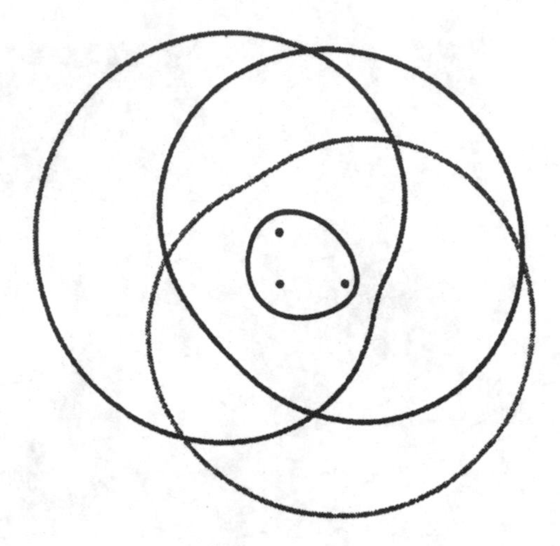

FIG. 4. *The Zariski closure of the 3-ellipse is an algebraic curve of degree eight.*

Figure 2 shows an algebraic 5-ellipse. In that picture, the actual 5-ellipse is the tiny convex curve in the center. It surrounds only one of the five foci.

For a less dizzying illustration see Figure 4. That picture shows an algebraic 3-ellipse. The curve has degree eight, and it is given algebraically by the 8×8-determinant displayed in the Introduction. We see that the set of real points on the algebraic 3-ellipse consists of four ovals, corresponding to the equations

$$\pm\sqrt{(x-u_1)^2+(y-v_1)^2} \pm \sqrt{(x-u_2)^2+(y-v_2)^2} \pm \sqrt{(x-u_3)^2+(y-v_3)^2} = d.$$

Thus Figure 4 visualizes the Galois theory argument in the proof of Lemma 2.1.

If we regard the radius d as an unknown, in addition to the two unknowns x and y, then the determinant in Theorem 1.1 specifies an irreducible surface $\{p_k(x, y, d) = 0\}$ in three-dimensional space. That surface has degree 2^k. For an algebraic geometer, this surface would live in complex projective 3-space $\mathbb{CP}^3$, but we are interested in its points in real affine 3-space $\mathbb{R}^3$. Figure 5 shows this surface for $k = 3$. The bowl-shaped convex branch near the top is the graph of the sum of distances function (3.1), while each of the other three branches is associated with a different combination of signs in the product (2.1). The surface has a total of $2^k = 8$ branches, but only the four in the half-space $d \geq 0$ are shown, as it is symmetric with respect to the plane $d = 0$. Note that the Fermat-Weber point (x_*, y_*, d_*) is a highly singular point of the surface.

The time has now come for us to explain the adjective "semidefinite" in the title of this paper. *Semidefinite programming* (SDP) is a widely used method in convex optimization. Introductory references include [17, 19].

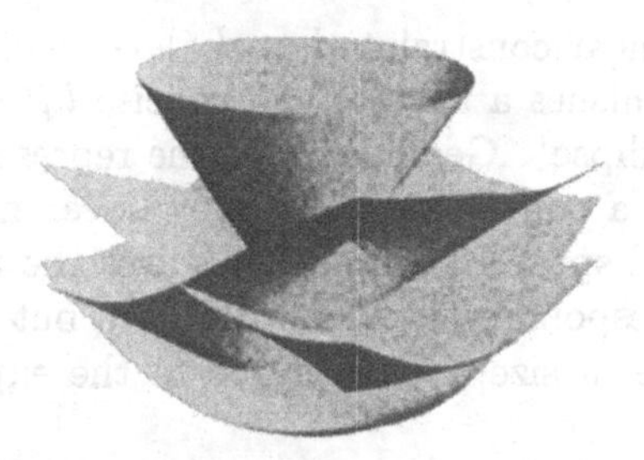

FIG. 5. *The irreducible surface* $\{p_3(x,y,d) = 0\}$. *Taking horizontal slices gives the pencil of algebraic 3-ellipses for fixed foci and different radii d, as shown in Figure 3.*

An algebraic perspective was recently given in [12]. The problem of SDP is to minimize a linear functional over the solution set of a linear matrix inequality (LMI). An example of an LMI is

$$x \cdot A_k + y \cdot B_k + d \cdot I_{2^k} + \tilde{C}_k \succeq 0. \tag{3.2}$$

Here $\tilde{C}_k$ is the matrix gotten from C_k by setting $d = 0$, so that $C_k = d \cdot I_{2^k} + \tilde{C}_k$. If d is a fixed positive real number then the solution set to the LMI (3.2) is the convex region $\mathcal{E}_k$ bounded by the k-ellipse. If d is an unknown then the solution set to the LMI (3.2) is the epigraph of (3.1), or, geometrically, the unbounded 3-dimensional convex region interior to the bowl-shaped surface in Figure 5. The bottom point of that convex region is the Fermat-Weber point (x_*, y_*, d_*), and it can be computed by solving the SDP

$$\text{Minimize } d \text{ subject to (3.2).} \tag{3.3}$$

Similarly, for fixed radius d, the k-ellipse is the set of all solutions to

$$\text{Minimize } \alpha x + \beta y \text{ subject to (3.2)} \tag{3.4}$$

where α, β run over $\mathbb{R}$. This explains the term *semidefinite representation* in our title.

While the Fermat-Weber SDP (3.3) has only three unknowns, it has the serious disadvantage that the matrices are exponentially large (size 2^k). For computing (x_*, y_*, d_*) in practice, it is better to introduce slack variables $d_1, d_2, \ldots, d_k$, and to solve

$$\text{Minimize } \sum_{i=1}^{k} d_i \text{ subject to } \begin{bmatrix} d_i + x - a_i & y - b_i \\ y - b_i & d_i - x + a_i \end{bmatrix} \succeq 0 \quad (i=1,\ldots,k). \tag{3.5}$$

This system can be written as a single LMI by stacking the 2×2-matrices to form a block matrix of size $2k \times 2k$. The size of the resulting LMI is linear in k while the size of the LMI (3.3) is exponential in k. If we take the LMI

(3.5) and add the linear constraint $d_1 + d_2 + \cdots + d_k = d$, for some fixed $d > 0$, then this furnishes a natural and concise *lifted* semidefinite representation of our k-ellipse[1]. Geometrically, the representation expresses $\mathcal{E}_k$ as the *projection* of a convex set defined by linear matrix inequalities in a higher-dimensional space. Theorem 1.1 solves the algebraic *LMI elimination problem* corresponding to this projection, but at the expense of an exponential increase in size, which is due to the exponential growing of degrees of k-ellipses.

Our last topic in this section is the relationship of the k-ellipse to the celebrated work of Helton and Vinnikov [5] on LMI representations of planar convex sets, which led to the resolution of the Lax conjecture in [8]. A semialgebraic set in the plane is called *rigidly convex* if its boundary has the property that every line passing through its interior intersects the Zariski closure of the boundary only in real points. Helton and Vinnikov [5, Thm. 2.2] proved that a plane curve of degree d has an LMI representation by symmetric $d \times d$ matrices if and only if the region bounded by this curve is rigidly convex. In arbitrary dimensions, rigid convexity holds for every region bounded by a hypersurface that is given by an LMI representation, but the strong form of the converse, where the degree of the hypersurface precisely matches the matrix size of the LMI, is only valid in two dimensions.

It follows from the LMI representation in Theorem 1.1 that the region bounded by a k-ellipse is rigidly convex. Rigid convexity can be seen in Figures 4 and 2. Every line that passes through the interior of the 3-ellipse intersects the algebraic 3-ellipse in eight real points, and lines through the 5-ellipse meet its Zariski closure in 32 real points. Combining our Theorem 1.1 with the Helton-Vinnikov Theorem, we conclude:

COROLLARY 3.1. *The k-ellipse is rigidly convex. If k is odd, it can be represented by an LMI of size 2^k, and if k is even, it can be represented as by LMI of size $2^k - \binom{k}{k/2}$.*

We have not found yet an *explicit* representation of size $2^k - \binom{k}{k/2}$ when k is even and $k \geq 4$. For the classical ellipse ($k = 2$), the determinantal representation (1.2) presented in the Introduction has size 4×4, while Corollary 3.1 guarantees the existence of a 2×2 representation. One such representation of the ellipse with foci (u_1, v_1) and (u_2, v_2) is:

$$\big(d^2 + (u_1 - u_2)(2x - u_1 - u_2) + (v_1 - v_2)(2y - v_1 - v_2)\big) \cdot I_2 + 2d \cdot \begin{bmatrix} x - u_2 & y - v_2 \\ y - v_2 & -x + u_2 \end{bmatrix} \succeq 0.$$

Notice that in this LMI representation of the ellipse, the matrix entries are linear in x and y, as required, but they are quadratic in the radius

[1]The formulation (3.5) actually provides a *second-order cone* representation of the k-ellipse. Second-order cone programming (SOCP) is another important class of convex optimization problems, of complexity roughly intermediate between that of linear and semidefinite programming; see [9] for a survey of the basic theory and its many applications.

parameter d and the foci u_i, v_i. What is the nicest generalization of this representation to the k-ellipse for k even?

4. Generalizations. The semidefinite representation of the k-ellipse we have found in Theorem 1.1 can be generalized in several directions. Our first generalization corresponds to the inclusion of arbitrary positive weights for the distances, while the second one extends the results from plane curves to higher dimensions. The resulting geometric shapes are known as *Tschirnhaus'sche Eiflächen* (or "egg surfaces") in the classical literature [11].

4.1. Weighted k-ellipse. Consider k points $(u_1, v_1), \ldots, (u_k, v_k)$ in the real plane $\mathbb{R}^2$, a positive radius d, and positive *weights* $w_1, \ldots, w_k$. The *weighted k-ellipse* is the plane curve defined as

$$\left\{ (x,y) \in \mathbb{R}^2 \; : \; \sum_{i=1}^{k} w_i \cdot \sqrt{(x-u_i)^2 + (y-v_i)^2} \; = \; d \right\},$$

where w_i indicates the relative weight of the distance from (x, y) to the i-th focus (u_i, v_i). The *algebraic weighted k-ellipse* is the Zariski closure of this curve. It is the zero set of an irreducible polynomial $p_k^w(x,y)$ that can be constructed as in equation (2.1). The interior of the weighted k-ellipse is the bounded convex region

$$\mathcal{E}_k(w) \; := \; \left\{ (x,y) \in \mathbb{R}^2 \; : \; \sum_{i=1}^{k} w_i \cdot \sqrt{(x-u_i)^2 + (y-v_i)^2} \; \le \; d \right\}.$$

The matrix construction from the unweighted case in (2.2) generalizes as follows:

$$L_k^w(x,y) := d \cdot I_{2^k} + w_1 \cdot \begin{bmatrix} x-u_1 & y-v_1 \\ y-v_1 & -x+u_1 \end{bmatrix} \oplus \cdots \oplus w_k \cdot \begin{bmatrix} x-u_k & y-v_k \\ y-v_k & -x+u_k \end{bmatrix}. \quad (4.1)$$

Each tensor summand is simply multiplied by the corresponding weight. The following representation theorem and degree formula are a direct generalization of Theorem 2.1:

THEOREM 4.1. *The algebraic weighted k-ellipse has the semidefinite representation*

$$p_k^w(x,y) \; = \; \det L_k^w(x,y),$$

and the convex region in its interior satisfies

$$\mathcal{E}_k(w) \; = \; \{ (x,y) \in \mathbb{R}^2 \; : \; L_k^w(x,y) \succeq 0 \}.$$

The degree of the weighted k-ellipse is given by

$$\deg p_k^w(x,y) \; = \; 2^k - |\mathcal{P}(w)|,$$

where $\mathcal{P}(w) = \{\delta \in \{-1,1\}^k : \sum_{i=1}^k \delta_i w_i = 0\}$.

Proof. The proof is entirely analogous to that of Theorem 2.1. □

A consequence of the characterization above is the following cute complexity result.

COROLLARY 4.1. *The decision problem "Given a weighted k-ellipse with fixed foci and positive integer weights, is its algebraic degree smaller than 2^k?" is NP-complete. The function problem "What is the algebraic degree?" is #P-hard.*

Proof. Since the number partitioning problem is NP-hard [4], the fact that our decision problem is NP-hard follows from the degree formula in Theorem 4.1. But it is also in NP because if the degree is less than 2^k, we can certify this by exhibiting a choice of δ_i for which $\sum_{i=1}^k \delta_i w_i = 0$. Computing the algebraic degree is equivalent to counting the number of solutions of a number partitioning problem, thus proving its #P-hardness. □

4.2. k-Ellipsoids. The definition of a k-ellipse in the plane can be naturally extended to a higher-dimensional space to obtain k-ellipsoids. Consider k fixed points $\mathbf{u}_1, \ldots, \mathbf{u}_k$ in $\mathbb{R}^n$, with $\mathbf{u}_i = (u_{i1}, u_{i2}, \ldots, u_{in})$. The *$k$-ellipsoid* in $\mathbb{R}^n$ with these foci is the hypersurface

$$\left\{\mathbf{x}\in\mathbb{R}^n : \sum_{i=1}^k \|\mathbf{u}_i - \mathbf{x}\| = d\right\} = \left\{\mathbf{x}\in\mathbb{R}^n : \sum_{i=1}^k \sqrt{\sum_{j=1}^n (u_{ij} - x_j)^2} = d\right\}. \quad (4.2)$$

This hypersurface encloses the convex region

$$\mathcal{E}_k^n \quad = \quad \left\{\mathbf{x} \in \mathbb{R}^n \; : \; \sum_{i=1}^k \|\mathbf{u}_i - \mathbf{x}\| \le d\right\}.$$

The Zariski closure of the k-ellipsoid is the hypersurface defined by an irreducible polynomial $p_k^n(\mathbf{x}) = p_k^n(x_1, x_2, \ldots, x_n)$. By the same reasoning as in Section 2, we can prove the following:

THEOREM 4.2. *The defining irreducible polynomial $p_n^k(\mathbf{x})$ of the k-ellipsoid is monic of degree 2^k in the parameter d, it has degree 2^k in $\mathbf{x}$ if k is odd, and it has degree $2^k - \binom{k}{k/2}$ if k is even.*

We shall represent the polynomial $p_n^k(\mathbf{x})$ as a factor of the determinant of a symmetric matrix of affine-linear forms. To construct this semidefinite representation of the k-ellipsoid, we proceed as follows. Fix an integer $m \ge 2$. Let $\mathbf{U}_i(\mathbf{x})$ be any symmetric $m\times m$-matrix of rank 2 whose entries are affine-linear forms in $\mathbf{x}$, and whose two non-zero eigenvalues are $\pm\|\mathbf{u}_i - \mathbf{x}\|$. Forming the tensor sum of these matrices, as in the proof of Theorem 2.1, we find that $p_k^n(\mathbf{x})$ is a factor of

$$\det\big(d \cdot I_{m^k} \; + \; \mathbf{U}_1(\mathbf{x}) \oplus \mathbf{U}_2(\mathbf{x}) \oplus \cdots \oplus \mathbf{U}_k(\mathbf{x})\big). \quad (4.3)$$

However, there are many extraneous factors. They are powers of the irreducible polynomials that define the k'-ellipsoids whose foci are subsets of $\{\mathbf{u}_1, \mathbf{u}_2, \dots, \mathbf{u}_k\}$.

There is a standard choice for the matrices $\mathbf{U}_i(\mathbf{x})$ that is symmetric with respect to permutations of the n coordinates. Namely, we can take $m = n+1$ and

$$\mathbf{U}_i(\mathbf{x}) = \begin{bmatrix} 0 & x_1 - u_{i1} & x_2 - u_{i2} & \cdots & x_n - u_{in} \\ x_1 - u_{i1} & 0 & 0 & \cdots & 0 \\ x_2 - u_{i2} & 0 & 0 & \cdots & 0 \\ \vdots & \vdots & \vdots & \ddots & \vdots \\ x_n - u_{in} & 0 & 0 & \cdots & 0 \end{bmatrix}.$$

However, in view of the extraneous factors in (4.3), it is desirable to replace these by matrices of smaller size m, possibly at the expense of having additional nonzero eigenvalues. It is not hard to see that $m = n$ is always possible, and the following result shows that, without extra factors, this is indeed the smallest possible matrix size.

LEMMA 4.1. *Let $A(x) := A_0 + A_1 x_1 + \cdots + A_n x_n$, where $A_0, A_1, \dots, A_n$ are real symmetric $m \times m$ matrices. If $\det A(x) = d^2 - \sum_{j=1}^n x_j^2$, then $m \geq n$.*

Proof. Assume, on the contrary, that $m < n$. For any fixed vector ξ with $\|\xi\| = d$ the matrix $A(\xi)$ must be singular. Therefore, there exists a nonzero vector $\eta \in \mathbb{R}^m$ such that $A(\xi)\eta = 0$. The set $\{x \in \mathbb{R}^n : A(x)\eta = 0\}$ is thus a nonempty affine subspace of positive dimension (at least $n - m$). The polynomial $d^2 - \sum_{j=1}^n x_j^2$ should then also vanish on this subspace, but this is not possible since $\{x \in \mathbb{R}^n : d^2 - \sum_{j=1}^n x_j^2 = 0\}$ is compact. □

However, if we allow extraneous factors or complex matrices, then the smallest possible value of m might drop. These questions are closely related to finding the *determinant complexity* of a given polynomial, as discussed in [10]. Note that in the applications to complexity theory considered there, the matrices of linear forms need not be symmetric.

5. Open questions and further research. The k-ellipse is an appealing example of an object from algebraic geometry. Its definition is elementary and intuitive, and yet it serves well in illustrating the intriguing interplay between algebraic concepts and convex optimization, in particular semidefinite programming. The developments presented in this paper motivate many natural questions. For most of these, to the best of our knowledge, we currently lack definite answers. Here is a short list of open problems and possible topics of future research.

Singularities and genus. Both the circle and the ellipse are rational curves, i.e., have genus zero. What is the genus of the (projective) algebraic k-ellipse? The first values, from $k = 1$ to $k = 4$, are 0,0,3,6. What is the formula for the genus in general? The genus is related to the class of the curve, i.e. the degree of the dual curve, and this number is the algebraic degree [12] of the problem (3.4). Moreover, is there a nice geometric characterization of all (complex) singular points of the algebraic k-ellipse?

Algebraic degree of the Fermat-Weber point. The Fermat-Weber point (x_*, y_*) is the unique solution of an algebraic optimization problem, formulated in (3.3) or (3.5), and hence it has a well-defined algebraic degree over $\mathbb{Q}(u_1, v_1, \ldots, u_k, v_k)$. However, that algebraic degree will depend on the combinatorial configuration of the foci. For instance, in the case $k = 4$ and foci forming a convex quadrilateral, the Fermat-Weber point lies in the intersection of the two diagonals [1], and therefore its algebraic degree is equal to one. What are the possible values for this degree? Perhaps a possible approach to this question would be to combine the results and formulas in [12] with the semidefinite characterizations obtained in this paper.

Reduced SDP representations of rigidly convex curves. A natural question motivated by our discussion in Section 3 is how to systematically produce minimal determinantal representations for a rigidly convex curve, when a non-minimal one is available. This is likely an easier task than finding a representation directly from the defining polynomial, since in this case we have a certificate of its rigid convexity.

Concretely, given real symmetric $n \times n$ matrices A and B such that

$$p(x, y) = \det(A \cdot x + B \cdot y + I_n)$$

is a polynomial of degree $r < n$, we want to produce $r \times r$ matrices $\tilde{A}$ and $\tilde{B}$ such that

$$p(x, y) = \det(\tilde{A} \cdot x + \tilde{B} \cdot y + I_r).$$

The existence of such matrices is guaranteed by the results in [5, 8]. In fact, explicit formulas in terms of theta functions of a Jacobian variety are presented in [5]. But isn't there a simpler algebraic construction in this special case?

Elimination in semidefinite programming. The projection of an algebraic variety is (up to Zariski closure, and over an algebraically closed field) again an algebraic variety. That projection can be computed using elimination theory or Gröbner bases. The projection of a polyhedron into a lower-dimensional subspace is a polyhedron. That projection can be computed using Fourier-Motzkin elimination. In contrast to these examples, the class of feasible sets of semidefinite programs is not closed under projections. As a simple concrete example, consider the convex set

$$\left\{(x, y, t) \in \mathbb{R}^3 \;:\; \begin{bmatrix} 1 & x-t \\ x-t & y \end{bmatrix} \succeq 0, \quad t \geq 0\right\}.$$

Its projection onto the (x, y)-plane is a convex set that is not rigidly convex, and hence cannot be expressed as $\{(x, y) \in \mathbb{R}^2 : Ax + By + C \succeq 0\}$.

In fact, that projection is not even basic semialgebraic. In some cases, however, this closure property nevertheless does hold. We saw this for the projection that transforms the representation (3.5) of the k-ellipse to the representation (3.3). Are there general conditions that ensure the semidefinite representability of the projections? Are there situations where the projection does not lead to an exponential blowup in the size of the representation?

Hypersurfaces defined by eigenvalue sums. Our construction of the (weighted) k-ellipsoid as the determinant of a tensor sum has the following natural generalization. Let $\mathbf{U}_1(\mathbf{x}), \mathbf{U}_2(\mathbf{x}), \ldots, \mathbf{U}_k(\mathbf{x})$ be any symmetric $m \times m$-matrices whose entries are affine-linear forms in $\mathbf{x} = (x_1, x_2, \ldots, x_n)$. Then we consider the polynomial

$$p(\mathbf{x}) \quad = \quad \det\big(d \cdot I_{m^k} + \mathbf{U}_1(\mathbf{x}) \oplus \mathbf{U}_2(\mathbf{x}) \oplus \cdots \oplus \mathbf{U}_k(\mathbf{x})\big). \qquad (5.1)$$

We also consider the corresponding rigidly convex set

$$\big\{\, \mathbf{x} \in \mathbb{R}^n \;:\; d \cdot I_{m^k} + \mathbf{U}_1(\mathbf{x}) \oplus \mathbf{U}_2(\mathbf{x}) \oplus \cdots \oplus \mathbf{U}_k(\mathbf{x}) \succeq 0 \,\big\}.$$

The boundary of this convex set is a hypersurface whose Zariski closure is the set of zeroes of the polynomial $p(\mathbf{x})$. It would be worthwhile to study the hypersurfaces of the special form (5.1) from the point of view of computational algebraic geometry.

These hypersurfaces specified by eigenvalue sums of symmetric matrices of linear forms have a natural generalization in terms of *resultant sums* of hyperbolic polynomials. For concreteness, let us take $k = 2$. If $p(\mathbf{x})$ and $q(\mathbf{x})$ are hyperbolic polynomials in n unknowns, with respect to a common direction $\mathbf{e}$ in $\mathbb{R}^n$, then the polynomial

$$(p \oplus q)(\mathbf{x}) \quad := \quad \mathrm{Res}_t\big(p(\mathbf{x} - t\mathbf{e}), q(\mathbf{x} + t\mathbf{e})\big)$$

is also hyperbolic with respect to $\mathbf{e}$. This construction mirrors the operation of taking Minkowski sums in the context of convex polyhedra, and we believe that it is fundamental for future studies of hyperbolicity in polynomial optimization [8, 13].

Acknowledgements. We are grateful to the IMA in Minneapolis for hosting us during our collaboration on this project. Bernd Sturmfels was partially supported by the U.S. National Science Foundation (DMS-0456960). Pablo A. Parrilo was partially supported by AFOSR MURI subaward 2003-07688-1 and the Singapore-MIT Alliance. We also thank the referee for his/her useful comments, and in particular for pointing out the #P-hardness part of Corollary 4.1.

REFERENCES

[1] C. Bajaj. The algebraic degree of geometric optimization problems. *Discrete Comput. Geom.*, **3**(2):177–191, 1988.

[2] R. Bellman. *Introduction to Matrix Analysis.* Society for Industrial and Applied Mathematics (SIAM), 1997.

[3] R.R. Chandrasekaran and A. Tamir. Algebraic optimization: the Fermat-Weber location problem. *Math. Programming*, **46**(2, (Ser. A)):219–224, 1990.

[4] M.R. Garey and D.S. Johnson. *Computers and Intractability: A guide to the theory of NP-completeness.* W.H. Freeman and Company, 1979.

[5] J.W. Helton and V. Vinnikov. Linear matrix inequality representation of sets. To appear in *Comm. Pure Appl. Math.* Preprint available from `arxiv.org/abs/math.OC/0306180`. 2003.

[6] R.A. Horn and C.R. Johnson. *Topics in Matrix Analysis.* Cambridge University Press, 1994.

[7] D.K. Kulshrestha. k-elliptic optimization for locational problems under constraints. *Operational Research Quarterly*, **28**(4-1):871–879, 1977.

[8] A.S. Lewis, P.A. Parrilo, and M.V. Ramana. The Lax conjecture is true. *Proc. Amer. Math. Soc.*, **133**(9):2495–2499, 2005.

[9] M. Lobo, L. Vandenberghe, S. Boyd, and H. Lebret. Applications of second-order cone programming. *Linear Algebra and its Applications*, **284**:193–228, 1998.

[10] T. Mignon and N. Ressayre. A quadratic bound for the determinant and permanent problem. *International Mathematics Research Notices*, **79**:4241–4253, 2004.

[11] G.Sz.-Nagy. Tschirnhaussche Eiflächen und Eikurven. *Acta Math. Acad. Sci. Hung.* **1**:36–45, 1950.

[12] J. Nie, K. Ranestad, and B. Sturmfels. The algebraic degree of semidefinite programming. Preprint, 2006, `math.OC/0611562`.

[13] J. Renegar. Hyperbolic programs, and their derivative relaxations. *Found. Comput. Math.* **6**(1):59–79, 2006.

[14] J. Sekino. n-ellipses and the minimum distance sum problem. *Amer. Math. Monthly*, **106**(3):193–202, 1999.

[15] H. Sturm. Über den Punkt kleinster Entfernungssumme von gegebenen Punkten. *Journal für die Reine und Angewandte Mathematik* **97**:49–61, 1884.

[16] C.M. Traub. *Topological Effects Related to Minimum Weight Steiner Triangulations.* PhD thesis, Washington University, 2006.

[17] L. Vandenberghe and S. Boyd. Semidefinite programming. *SIAM Review*, **38**: 49-95, 1996.

[18] E.V. Weiszfeld. Sur le point pour lequel la somme des distances de n points donnés est minimum. *Tohoku Mathematical Journal* **43**:355–386, 1937.

[19] H. Wolkowicz, R. Saigal, and L. Vandenberghe (Eds.). *Handbook of Semidefinite Programming.* Theory, Algorithms, and Applications Series: International Series in Operations Research and Management Science, Vol. **27**, Springer Verlag, 2000.

SOLVING POLYNOMIAL SYSTEMS EQUATION BY EQUATION

ANDREW J. SOMMESE*, JAN VERSCHELDE†, AND CHARLES W. WAMPLER‡

Abstract. By a numerical continuation method called a diagonal homotopy, one can compute the intersection of two irreducible positive dimensional solution sets of polynomial systems. This paper proposes to use this diagonal homotopy as the key step in a procedure to intersect general solution sets that are not necessarily irreducible or even equidimensional. Of particular interest is the special case where one of the sets is defined by a single polynomial equation. This leads to an algorithm for finding a numerical representation of the solution set of a system of polynomial equations introducing the equations one by one. Preliminary computational experiments show this approach can exploit the special structure of a polynomial system, which improves the performance of the path following algorithms.

Key words. Algebraic set, component of solutions, diagonal homotopy, embedding, equation-by-equation solver, generic point, homotopy continuation, irreducible component, numerical irreducible decomposition, numerical algebraic geometry, path following, polynomial system, witness point, witness set.

AMS(MOS) subject classifications. Primary 65H10; Secondary 13P05, 14Q99, 68W30.

1. Introduction. Homotopy continuation methods provide reliable and efficient numerical algorithms to compute accurate approximations to all isolated solutions of polynomial systems, see e.g. [16] for a recent survey. As proposed in [28], we can approximate a positive dimensional solution set of a polynomial system by isolated solutions, which are obtained as intersection points of the set with a generic linear space of complementary dimension.

---

*Department of Mathematics, University of Notre Dame, Notre Dame, IN 46556-4618, USA (sommese@nd.edu; www.nd.edu/~sommese). This material is based upon work supported by the National Science Foundation under Grant No. 0105653 and Grant No. 0410047; the Duncan Chair of the University of Notre Dame; the Land Baden-Württemberg (RiP-program at Oberwolfach); and the Institute for Mathematics and Its Applications (IMA), Minneapolis.

†Department of Mathematics, Statistics, and Computer Science, University of Illinois at Chicago, 851 South Morgan (M/C 249), Chicago, IL 60607-7045, USA (jan@math.uic.edu or jan.verschelde@na-net.ornl.gov; www.math.uic.edu/~jan). This material is based upon work supported by the National Science Foundation under Grant No. 0105739, Grant No. 0134611, and Grant No. 0410036; the Land Baden-Württemberg (RiP-program at Oberwolfach); and the Institute for Mathematics and Its Applications (IMA), Minneapolis

‡General Motors Research and Development, Mail Code 480-106-359, 30500 Mound Road, Warren, MI 48090-9055, U.S.A. (Charles.W.Wampler@gm.com). This material is based upon work supported by the National Science Foundation under Grant No. 0410047; General Motors Research and Development; the Land Baden-Württemberg (RiP-program at Oberwolfach); and the Institute for Mathematics and Its Applications (IMA), Minneapolis.

New homotopy algorithms have been developed in a series of papers [19, 20, 23, 26, 27] to give numerical representations of positive dimensional solution sets of polynomial systems. These homotopies are the main numerical algorithms in a young field we call *numerical algebraic geometry*. See [29] for a detailed treatment of this subject.

This paper provides an algorithm to compute numerical approximations to positive dimensional solution sets of polynomial systems by introducing the equations a few at a time, i.e., "subsystem by subsystem," or even one at a time, i.e., "equation by equation." The advantage of working in this manner is that the special properties of subsets of the equations are revealed early in the process, thus reducing the computational cost of later stages. Consequently, although the new algorithm has more stages of computation than earlier approaches, the amount of work in each stage can be considerably less, producing a net savings in computing time.

This paper is organized in three parts. In §2, we explain our method to represent and to compute a numerical irreducible decomposition of the solution set of a polynomial system. In §2.3, we relate our numerical irreducible decomposition to the symbolic geometric resolution. In §3, the new diagonal homotopy algorithm is applied to solve systems subsystem by subsystem or equation by equation. Computational experiments are given in §4.

2. A numerical irreducible decomposition.

We start this section with a motivating illustrative example that shows the occurrence of several solution sets of different dimensions and degrees. Secondly, we define the notion of witness sets, which represent pure dimensional solution sets of polynomial systems *numerically*. Witness sets are computed by cascades of homotopies between embeddings of polynomial systems.

2.1. An illustrative example.

Our running example (used also in [20]) is the following:

$$f(x,y,z) = \left[\begin{array}{r} (y-x^2)(x^2+y^2+z^2-1)(x-0.5) \\ (z-x^3)(x^2+y^2+z^2-1)(y-0.5) \\ (y-x^2)(z-x^3)(x^2+y^2+z^2-1)(z-0.5) \end{array}\right]. \quad (2.1)$$

In this factored form we can easily identify the decomposition of the solution set $Z = f^{-1}(\mathbf{0})$ into irreducible solution components, as follows:

$$Z = Z_2 \cup Z_1 \cup Z_0 = \{Z_{21}\} \cup \{Z_{11} \cup Z_{12} \cup Z_{13} \cup Z_{14}\} \cup \{Z_{01}\} \quad (2.2)$$

where

1. Z_{21} is the sphere $x^2+y^2+z^2-1=0$,
2. Z_{11} is the line $(x=0.5, z=0.5^3)$,
3. Z_{12} is the line $(x=\sqrt{0.5}, y=0.5)$,
4. Z_{13} is the line $(x=-\sqrt{0.5}, y=0.5)$,
5. Z_{14} is the twisted cubic $(y-x^2=0, z-x^3=0)$,
6. Z_{01} is the point $(x=0.5, y=0.5, z=0.5)$.

The sequence of homotopies in [19] tracks 197 paths to find a numerical representation of the solution set Z. With the new approach we have to trace only 13 paths! We show how this is done in Figure 3 in §4.1 below, but we first describe a numerical representation of Z in the next section.

2.2. Witness sets. We define witness sets as follows. Let $f : \mathbb{C}^N \rightarrow \mathbb{C}^n$ define a system $f(\mathbf{x}) = \mathbf{0}$ of n polynomial equations $f = \{f_1, f_2, \ldots, f_n\}$ in N unknowns $\mathbf{x} = (x_1, x_2, \ldots, x_N)$. We denote the solution set of f by

$$V(f) = \{ \mathbf{x} \in \mathbb{C}^N \mid f(\mathbf{x}) = \mathbf{0} \}. \tag{2.3}$$

This is a reduced[1] algebraic set. Suppose $X \subset V(f) \subset \mathbb{C}^N$ is a pure dimensional[2] algebraic set of dimension i and degree d_X. Then, a witness set for X is a data structure consisting of the system f, a generic linear space $L \subset \mathbb{C}^N$ of codimension i, and the set of d_X points $X \cap L$.

If X is not pure dimensional, then a witness set for X breaks up into a list of witness sets, one for each dimension. In our work, we generally ignore multiplicities, so when a polynomial system has a nonreduced solution component, we compute a witness set for the reduction of the component. Just as X has a unique decomposition into irreducible components, a witness set for X has a decomposition into the corresponding irreducible witness sets, represented by a partition of the witness set representation for X. We call this a *numerical irreducible decomposition* of X, introduced in [20]..

The irreducible decomposition of the solution set Z in (2.2) is represented by

$$[W_2, W_1, W_0] = [[W_{21}], [W_{11}, W_{12}, W_{13}, W_{14}], [W_{01}]], \tag{2.4}$$

where the W_i are witness sets for pure dimensional components, of dimension i, partitioned into witness sets W_{ij}'s corresponding to the irreducible components of Z. In particular:

1. W_{21} contains two points on the sphere, cut out by a random line;
2. W_{11} contains one point on the line $(x = 0.5, z = 0.5^3)$, cut out by a random plane;
3. W_{12} contains one point on the line $(x = \sqrt{0.5}, y = 0.5)$, cut out by a random plane;
4. W_{13} contains one point on the line $(x = -\sqrt{0.5}, y = 0.5)$, cut out by a random plane;
5. W_{14} contains three points on the twisted cubic, cut out by a random plane; and
6. W_{01} is just the point $(x = 0.5, y = 0.5, z = 0.5)$.

[1] "Reduced" means the set occurs with multiplicity one, we ignore multiplicities > 1 in this paper.

[2] "Pure dimensional" (or "equidimensional") means all components of the set have the same dimension.

Applying the formal definition, the witness sets W_{ij} consist of witness points $\mathbf{w} = \{i, f, L, \mathbf{x}\}$, for $\mathbf{x} \in Z_{ij} \cap L$, where L is a random linear subspace of codimension i (in this case, of dimension $3 - i$). Moreover, observe $\#W_{ij} = \deg(Z_{ij}) = \#(Z_{ij} \cap L)$.

2.3. Geometric resolutions and triangular representations. Witness sets are slices by generic linear spaces of complementary dimension. Linear space sections have been studied in algebraic geometry since the nineteenth century, e.g., see [1]. There has also been considerable modern interest in commutative algebra in the properties of slices by generic linear spaces of complementary dimension, e.g., [3, 9].

Such linear slices are used in symbolic calculations under the name of *lifting fibers* which occur in a *geometric resolution* of polynomial system. The geometric resolution is a symbolic analogue to a numerical irreducible decomposition and was introduced in the symbolic community independently of the introduction in the numerical community. The origins of the numerical development go back to [28], while [4] appears to be foundational for the symbolic approach. As indicated in the title of [4], the dimension and the isolated points of a variety can be computed in polynomial time. Avoiding multivariate polynomials to describe positive dimensional solution sets and using straight-line programs, the notion of geometric resolution was introduced in [6] and lifting fibers were further developed in [5], [7], and [15], leading to a "Gröbner free algorithm" to solve polynomial systems.

TABLE 1
A dictionary between lifting fibers and witness sets.

Generic Points on a Pure Dimensional Solution Set V	
Symbolic: lifting fiber $\pi^{-1}(\mathbf{p})$	Numeric: witness set $W(\mathbf{c})$
computational field k: numbers in $\mathbb{Q}$ (or a finite extension) field operations done *symbolically*	*numeric* field $\mathbb{C}$: floating point complex numbers with machine arithmetic
With a *symbolic* coordinate change we bring V to Noether position: replace $\mathbf{x}$ by $M\mathbf{y}$, $M \in k^{n \times n}$	We slice V *numerically* with some randomly chosen hyperplanes: $A\mathbf{x} = \mathbf{c}$, $A \in \mathbb{C}^{r \times n}$, $\mathbf{c} \in \mathbb{C}^r$, $\mathrm{rank}(A) = r$
choose M for coordinate change	*choose* A for slicing hyperplanes
$\dim V = r$: specialize r free variables	$\dim V = r$: cut with r hyperplanes
$\pi^{-1}(\mathbf{p}) = \{\ \mathbf{y} \in \mathbb{C}^n \mid f(\mathbf{y}) = 0$ and $y_1 = p_1, \dots, y_r = p_r\ \}$	$W(\mathbf{c}) = \{\ \mathbf{x} \in \mathbb{C}^n \mid$ $f(\mathbf{x}) = 0$ and $A\mathbf{x} = \mathbf{c}\ \}$
choice of values $\mathbf{p} = (p_1, \dots, p_r)$ for free variables $(y_1, \dots, y_r)$ such that the fiber $\pi^{-1}(\mathbf{p})$ is finite	*choice* of r constants $\mathbf{c} = (c_1, \dots, c_r)$ so that $\left\{ \begin{array}{c} f(\mathbf{x}) = 0 \\ A\mathbf{x} = \mathbf{c} \end{array} \right.$ has isolated solutions
for almost all $\mathbf{p} \in k^r$: $\pi^{-1}(\mathbf{p})$ consists of $\deg V$ smooth points	*for almost all* $\mathbf{c} \in \mathbb{C}^r$: $W(\mathbf{c})$ consists of $\deg V$ smooth points
for almost all means except for a proper algebraic subset of bad choices.	

Table 1 lists a dictionary relating symbolic lifting fibers and numeric witness sets for representing a pure dimensional solution set V of dimension r. Let us first look at the similarities between the two data structures: by random coordinate changes followed by a specialization of r variables or by adding r linear equations with random coefficients, we obtain as many generic points as the degree of the set V. (To emphasize the correspondences in the table, we have defined witness sets as the intersection with extrinsically-specified hyperplanes, $A\mathbf{x} = \mathbf{c}$, $\text{rank}(A) = r$, but for efficiency, we may work instead with intrinsically-specified linear spaces, $\mathbf{x} = B\mathbf{y} + b$, $\mathbf{y} \in \mathbb{C}^{n-r}$, see [27, 28].) For almost all choices of random numbers, the lifting fibers and witness sets consist of $\deg(V)$ many regular points.

A major difference between the two data structures lies in the choice of number field. When working symbolically, exact arithmetic with rational numbers is used, whereas numerically, the calculations are performed in complex floating point arithmetic. Another significant difference is the form of the final answer. Symbolic solutions are given implicitly by a new set of algebraic relations having a specified form, for example, the result might be represented as a polynomial in one variable, say $p(u) \in k[u]$, such that the points in the lifting fiber are given by $\mathbf{x} = \mathbf{S}(u)$ for some straight-line program $\mathbf{S}(u)$ evaluated at the $\deg(p)$ roots of $p(u)$ (see [5]). The output of the symbolic algorithm is $p(u)$ and $\mathbf{S}(u)$. In contrast, a numerical witness set consists of a list of floating point approximations to the witness points. While mathematically, lifting fibers and witness sets are geometrically equivalent descriptions of a pure dimensional solution set, in practice, the choice of number field and the form of the representation of the solution dramatically affect the kind of algorithms that are used and the types of applications that can be solved.

Triangular representations are an alternative to geometric resolutions for symbolically representing positive dimensional solutions sets of polynomial systems. In [14], Lazard defines triangular sets, a refinement of the characteristic sets of Ritt and Wu. Given a polynomial system, the algorithm in [14] returns a list of triangular sets, each of which represents a "quasi-component." (A quasi-component is a pure dimensional, not necessarily irreducible, solution component.) In [13], Kalkbrenner gives an algorithm to compute regular chains, a related type of triangular representation. Both of these approaches proceed equation by equation, producing triangular representations for the solution set at each stage. As in our approach, inclusion tests are needed to remove redundant components. Kalkbrenner notes that regular chains represent pure dimensional components without factoring them into irreducible components, which yields a savings in computation. Our numerical approach also first computes the pure dimensional components, leaving their factorization into irreducibles as an optional post-processing step. While these approaches resemble ours in the superficial respects of proceeding equation by equation and in eliminating redundancies by inclusion tests, it should be appreciated that the underlying processes are completely different. The computation of triangular sets for systems with approximate input coefficients is addressed in [17, 18].

2.4. Embeddings and cascades of homotopies. A witness superset $\widehat{W}_k$ for the pure k-dimensional part X_k of X is a set in $X \cap L$, which contains $W_k := X_k \cap L$ for a generic linear space L of codimension k. The set of "junk points" in $\widehat{W}_k$ is the set $\widehat{W}_k \setminus W_k$, which lies in $(\cup_{j>k} X_j) \cap L$.

The computation of a numerical irreducible decomposition for X runs in three stages:

1. Computation of a *witness superset* $\widehat{W}$ consisting of witness supersets $\widehat{W}_k$ for each dimension $k = 0, 1, 2, \ldots, N$.
2. Removal of junk points from $\widehat{W}$ to get a witness set W for X.
3. Decomposition of W into its irreducible components by partitioning each witness set W_k into witness sets corresponding to the irreducible components of X_k.

Up to this point, we have used the dimension of a component as the subscript for its witness set, but in the algorithms that follow, it will be more convenient to use codimension. The original algorithm for constructing witness supersets was given in [28]. A more efficient cascade algorithm for this was given in [19] by means of an embedding theorem.

In [26], we showed how to carry out the generalization of [19] to solve a system of polynomials on a pure N-dimensional algebraic set $Z \subset \mathbb{C}^m$. In the same paper, we used this capability to address the situation where we have two polynomial systems f and g on $\mathbb{C}^N$ and we wish to describe the irreducible decompositions of $A \cap B$ where $A \in \mathbb{C}^N$ is an irreducible component of $V(f)$ and $B \in \mathbb{C}^N$ is an irreducible component of $V(g)$. We call the resulting algorithm a *diagonal homotopy*, because it works by decomposing the diagonal system $\mathbf{u} - \mathbf{v} = \mathbf{0}$ on $Z = A \times B$, where $(\mathbf{u}, \mathbf{v}) \in \mathbb{C}^{2N}$. In [27], we rewrote the homotopies "intrinsically," which means that the linear slicing subspaces are not described explicitly by linear equations vanishing on them, but rather by linear parameterizations. (Note that intrinsic forms were first used in a substantial way to deal with numerical homotopies of parameterized linear spaces in [11], see also [12].) This has always been allowed, even in [28], but [27] showed how to do so consistently through the cascade down dimensions of the diagonal homotopy, thereby increasing efficiency by using fewer variables.

The subsequent steps of removing junk and decomposing the witness sets into irreducible pieces have been studied in [20, 21, 22, 23]. These methods presume the capability to track witness points on a component as the linear slicing space is varied continuously. This is straightforward for reduced solution components, but the case of nonreduced components, treated in [24], is more difficult. An extended discussion of the basic theory may be found in [29].

In this paper, we use multiple applications of the diagonal homotopy to numerically compute the irreducible decomposition of $A \cap B$ for general algebraic sets A and B, without the restriction that they be irreducible. At first blush, this may seem an incremental advance, basically consisting of organizing the requisite bookkeeping without introducing any significantly new theoretical constructs. However, this approach becomes particularly interesting when it is applied "equation by equation," that is, when we compute the irreducible decomposition of $V(f)$ for a system $f = \{f_1, f_2, \ldots, f_n\}$ by systematically computing $V(f_1)$, then $A_1 \cap V(f_2)$ for A_1 a component of $V(f_1)$, then $A_2 \cap V(f_3)$ for A_2 a component of $A_1 \cap V(f_2)$, etc. In this way, we incrementally build up the irreducible decomposition one equation at a time, by intersecting the associated hypersurface with all the solution components of the preceding equations. The main impact is that the elimination of junk points and degenerate solutions at early stages in the computation streamlines the subsequent stages. Even though we use only the total degree of the equations—not multihomogeneous degrees or Newton polytopes—the approach is surprisingly effective for finding isolated solutions.

3. Application of diagonal homotopies. In this section, we define our new algorithms by means of two flowcharts, one for solving subsystem by subsystem, and one that specializes the first one to solving equation by equation. We then briefly outline simplifications that apply in the case that only the nonsingular solutions are wanted. First, though, we summarize the notation used in the definition of the algorithms.

3.1. Symbols used in the algorithms. A witness set W for a pure i-dimensional component X in $V(f)$ is of the form $W = \{i, f, L, \mathcal{X}\}$, where L is the linear subspace that cuts out the $\deg X$ points $\mathcal{X} = X \cap L$. In the following algorithm, when we speak of a *witness point* $\mathbf{w} \in W$, it means that $\mathbf{w} = \{i, f, L, \mathbf{x}\}$ for some $\mathbf{x} \in \mathcal{X}$. For such a $\mathbf{w}$ and for g a polynomial (system) on $\mathbb{C}^N$, we use the shorthand $g(\mathbf{w})$ to mean $g(\mathbf{x})$, for $\mathbf{x} \in \mathbf{w}$.

In analogy to $V(f)$, which acts on a polynomial system, we introduce the operator $\mathcal{V}(W)$, which means the solution component represented by the witness set W. We also use the same symbol operating on a single witness point $\mathbf{w} = \{i, f, L, \mathbf{x}\}$, in which case $\mathcal{V}(\mathbf{w})$ means the irreducible component of $V(f)$ on which point $\mathbf{x}$ lies. This is consistent in that $\mathcal{V}(W)$ is the union of $\mathcal{V}(\mathbf{w})$ for all $\mathbf{w} \in W$.

Another notational convenience is the operator $\mathcal{W}(A)$, which gives a witness set for an algebraic set A. This is not unique, as it depends on the choice of the linear subspaces that slice out the witness points. However, any two witness sets $W_1, W_2 \in \mathcal{W}(A)$ are equivalent under a homotopy that smoothly moves from one set of slicing subspaces to the other, avoiding a proper algebraic subset of the associated Grassmannian spaces, where witness points diverge or cross. That is, we have $\mathcal{V}(\mathcal{W}(A)) = A$ and $\mathcal{W}(\mathcal{V}(W)) \equiv W$, where the equivalence in the second expression is under homotopy continuation between linear subspaces.

The output of our algorithm is a collection of witness sets $W_{\langle i \rangle}$, $i = 1, 2, \ldots, N$, where $W_{\langle i \rangle}$ is a witness set for the pure *codimension* i component of $V(f)$. (Note that the angle brackets, $\langle\,\rangle$, in the subscript distinguishes from our usual convention of subscripting by dimension. We

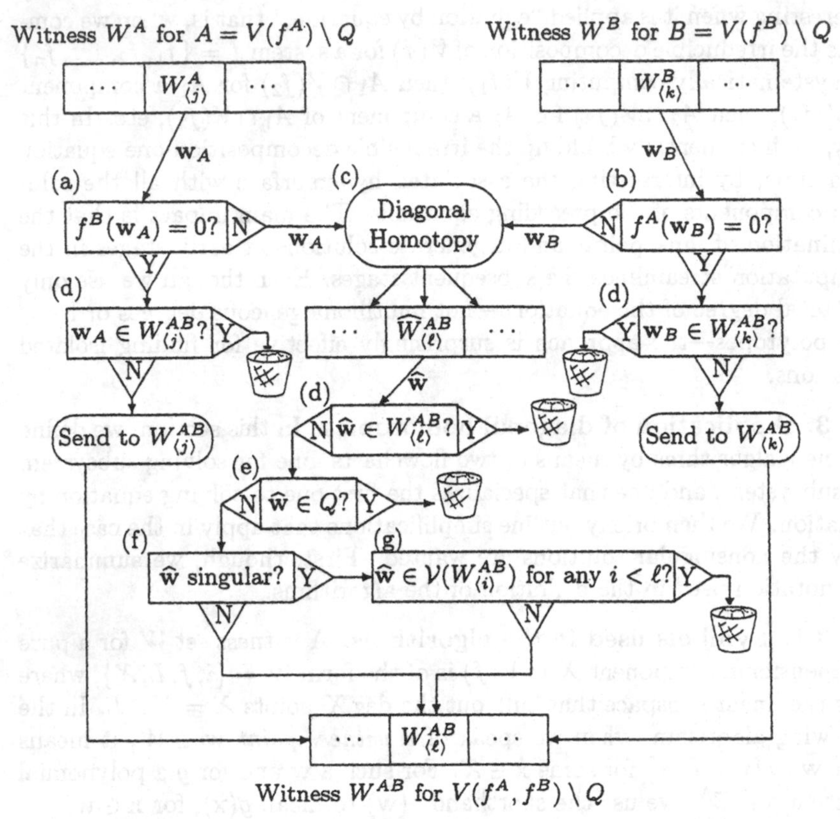

FIG. 1. *Subsystem-by-subsystem generation of witness sets for* $V(f^A, f^B) \setminus Q$.

introduce this notation because codimension is more natural for our current algorithm.) Breaking $W_{\langle i \rangle}$ into irreducible pieces is a post-processing task, done by techniques described in [20, 22, 23], which will not be described here.

The algorithm allows the specification of an algebraic set $Q \in \mathbb{C}^N$ that we wish to ignore. That is, we drop from the output any components that are contained in Q, yielding witness sets for $V(f_1, f_2, \cdots, f_n) \in \mathbb{C}^N \setminus Q$. Set Q can be specified as a collection of polynomials defining it or as a witness point set.

For convenience, we list again the operators used in our notation, as follows:

$V(f)$ The solution set of $f(x) = 0$.
$\mathcal{W}(A)$ A witness set for an algebraic set A, multiplicities ignored, as always.
$\mathcal{V}(W)$ The solution component represented by witness set W.
$\mathcal{V}(\mathbf{w})$ The irreducible component of $V(f)$ on which witness point $\mathbf{w} \in \mathcal{W}(V(f))$ lies.

3.2. Solving subsystem by subsystem. In this section, we describe how the diagonal homotopy can be employed to generate a witness set $W = \mathcal{W}(V(f^A, f^B) \setminus Q)$, given witness sets W^A for $A = V(f^A) \setminus Q$, and W^B for $B = V(f^B) \setminus Q$. Let us denote this operation as $W =$ **SysBySys**$(A, B; Q)$. Moreover, suppose **Witness**$(f; Q)$ computes a witness set $\mathcal{W}(V(f) \setminus Q)$ by any means available, such as by working on the entire system f as in our previous works, [19, 28], with junk points removed but not necessarily decomposing the sets into irreducibles. With these two operations in hand, one can approach the solution of any large system of polynomials in stages. For example, suppose $f = \{f^A, f^B, f^C\}$ is a system of polynomials composed of three subsystems, f^A, f^B, and f^C, each of which is a collection of one or more polynomials. The computation of a witness set $W = \mathcal{W}(V(f) \setminus Q)$ can be accomplished as

$$\begin{aligned} W^A &= \textbf{Witness}(f^A; Q), \quad W^B = \textbf{Witness}(f^B; Q) \\ W^C &= \textbf{Witness}(f^C; Q), \quad W^{AB} = \textbf{SysBySys}(W^A, W^B; Q), \\ &\quad W = \textbf{SysBySys}(W^{AB}, W^C; Q). \end{aligned}$$

This generalizes in an obvious way to any number of subsystems. Although we could compute $W =$ **Witness**$(f; Q)$ by directly working on the whole system f in one stage, there can be advantages to breaking the computation into smaller stages.

The diagonal homotopy as presented in [26] applies to computing $A \cap B$ only when A and B are each irreducible. To implement **SysBySys**, we need to handle sets that have more than one irreducible piece. In simplest terms, the removal of the requirement of irreducibility merely entails looping through all pairings of the irreducible pieces of A and B, followed by filtering to remove from the output any set that is contained inside another set in the output, or if two sets are equal, to eliminate the duplication. In addition to this, however, we would like to be able to proceed without first decomposing A and B into irreducibles. With a bit of attention to the details, this can be arranged.

Figure 1 gives a flowchart for algorithm **SysBySys**. For this to be valid as shown, we require that the linear subspaces for slicing out witness sets are chosen once and for all and used in all the runs of **Witness** and **SysBySys**. In other words, the slicing subspaces for W^A and W^B at the top of the algorithm must be the same as each other and as the output W^{AB}. This ensures that witness sets from one stage can, under certain circumstances, pass directly through to the next stage. Otherwise, a continuation step would need to be inserted to move from one slicing subspace to another.

The setup of a diagonal homotopy to intersect two irreducibles $A \subset \mathbb{C}^N$ and $B \subset \mathbb{C}^N$ involves the selection of certain random elements. We refer to [26, 27] for the full details. All we need to know at present is that in choosing these random elements the only dependence on A and B is their dimensions, $\dim A$ and $\dim B$. If we were to intersect another pair of irreducibles, say $A' \subset \mathbb{C}^N$ and $B' \subset \mathbb{C}^N$, having the same dimensions as the first pair, i.e., $\dim A' = \dim A$ and $\dim B' = \dim B$, then we may use the same random elements for both. In fact, the random choices will be generic for any finite number of intersection pairs. Furthermore, if A and A' are irreducible components of the solution set of the same system of polynomials, f^A, and B and B' are similarly associated to system f^B, then we may use exactly the same diagonal homotopy to compute $A \cap B$ and $A' \cap B'$. The only difference is that in the former case, the start points of the homotopy are pairs of points $(\alpha, \beta) \in \mathcal{W}(A) \times \mathcal{W}(B) \subset \mathbb{C}^{2N}$, while in the latter, the start points come from $\mathcal{W}(A') \times \mathcal{W}(B')$.

To explain this more explicitly, consider that the diagonal homotopy for intersecting A with B works by decomposing $\mathbf{u} - \mathbf{v}$ on $A \times B$. To set up the homotopy, we form the randomized system

$$\mathcal{F}(\mathbf{u}, \mathbf{v}) = \begin{bmatrix} R_A f^A(\mathbf{u}) \\ R_B f^B(\mathbf{v}) \end{bmatrix}, \tag{3.1}$$

where R_A is a random matrix of size $(N - \dim A) \times \#(f^A)$ and R_B is random of size $(N - \dim B) \times \#(f^B)$. [By $\#(f^A)$ we mean the number of polynomials in system f^A and similarly for $\#(f^B)$.] The key property is that $A \times B$ is an irreducible component of $V(\mathcal{F}(\mathbf{u}, \mathbf{v}))$ for all (R_A, R_B) in a nonzero Zariski open subset of $\mathbb{C}^{(N-\dim A)\times\#(f^A)} \times \mathbb{C}^{(N-\dim B)\times\#(f^B)}$, say R_{AB}. But this property holds for $A' \times B'$ as well, on a possibly different Zariski open subset, say $R_{A'B'}$. But $R_{AB} \cap R_{A'B'}$ is still a nonzero Zariski open subset, that is, almost any choice of (R_A, R_B) is satisfactory for computing both $A \cap B$ and $A' \cap B'$, and by the same logic, for any finite number of such intersecting pairs.

The upshot of this is that if we wish to intersect a pure dimensional set $A = \{A_1, A_2\} \subset V(f^A)$ with a pure dimensional set $B = \{B_1, B_2\} \subset V(f^B)$, where A_1, A_2, B_1, and B_2 are all irreducible, we may form one diagonal homotopy to compute all four intersections $A_i \cap B_j$, $i, j \in \{1, 2\}$, feeding in start point pairs from all four pairings. In short, the algorithm is completely indifferent as to whether A and B are irreducible or not. Of course, it can happen that the same irreducible component of $A \cap B$ can arise from more than one pairing $A_i \cap B_j$, so we will need to take steps to eliminate such duplications.

We are now ready to examine the details of the flowchart in Figure 1 for computing $W^{AB} = \mathcal{W}(V(f^A, f^B) \setminus Q)$ from $W^A = \mathcal{W}(V(f^A) \setminus Q)$ and $W^B = \mathcal{W}(V(f^B) \setminus Q)$. It is assumed that the linear slicing subspaces are the same for W^A, W^B, and W^{AB}. The following items (a)–(g) refer to labels in that chart.

(a) Witness point $\mathbf{w}_A$ is a generic point of the component of $V(f^A)$ on which it lies, $\mathcal{V}(\mathbf{w}_A)$. Consequently, $f^B(\mathbf{w}_A) = 0$ implies, with probability one, that $\mathcal{V}(\mathbf{w}_A)$ is contained in some component of $V(f^B)$. Moreover, we already know that $\mathbf{w}_A$ is not in any higher dimensional set of A, and therefore it cannot be in any higher dimensional set of $A \cap B$. Accordingly, any point $\mathbf{w}_A$ that passes test (a) is an isolated point in witness superset $\widehat{W}^{AB}$. The containment of $\mathcal{V}(\mathbf{w}_A)$ in B means that the dimension of the set is unchanged by intersection, so if $\mathbf{w}_A$ is drawn from $W^A_{\langle j \rangle}$, its correct destination is $W^{AB}_{\langle j \rangle}$.

On the other hand, if $f^B(\mathbf{w}_A) \neq 0$, then $\mathbf{w}_A$ proceeds to the diagonal homotopy as part of the computation of $\mathcal{V}(\mathbf{w}_A) \cap B$.

(b) This is the symmetric operation to (a).

(c) Witness points for components not completely contained in the opposing system are fed to the diagonal homotopy in order to find the intersection of those components. For each combination (a, b), where $a = \dim \mathcal{V}(\mathbf{w}_A)$ and $b = \dim \mathcal{V}(\mathbf{w}_B)$, there is a diagonal homotopy whose random constants are chosen once and for all at the start of the computation.

(d) This test, which appears in three places, makes sure that multiple copies of a witness point do not make it into W^{AB}. Such duplications can arise when A and B have components in common, when different pairs of irreducible components from A and B share a common intersection component, or when some component is nonreduced.

(e) Since a witness point $\widehat{\mathbf{w}}$ is sliced out generically from the irreducible component, $\mathcal{V}(\widehat{\mathbf{w}})$, on which it lies, if $\widehat{\mathbf{w}} \in Q$, then $\mathcal{V}(\widehat{\mathbf{w}}) \subset Q$. We have specified at the start that we wish to ignore such sets, so we throw them out here.

(f) In this test, "singular" means that the Jacobian matrix of partial derivatives for the sliced system that cuts out the witness point is rank deficient. We test this by a singular value decomposition of the matrix. If the point is nonsingular, it must be isolated and so it is clearly a witness point. On the other hand, if it is singular, it might be either a singular isolated point or it might be a junk point that lies on a higher dimensional solution set, so it must be subjected to further testing.

(g) Our current test for whether a singular test point is isolated or not is to check it against all the higher dimensional sets. If it is not in any of these, then it must be an isolated point, and we put it in the appropriate output bin.

In the current state of the art, the test in box (g) is done using homotopy membership tests. This consists of following the paths of the witness points of the higher dimensional set as its linear slicing subspace is moved continuously to a generically disposed one passing through the test point. The test point is in the higher dimensional set if, and only if, at the end of this continuation one of these paths terminates at the test point, see [21]. In the future, it may be possible that a reliable local test, based just on the local behavior of the polynomial system, can be devised that determines if a point is isolated or not. This might substantially reduce the computation required for the test. As it stands, one must test the point against all higher dimensional solution components, and so points reaching box (g) may have to wait there in limbo until all higher dimensional components have been found.

The test (e) for membership in Q would entail a homotopy membership test if Q is given by a witness set. If Q is given as $V(f^Q)$ for some polynomial system f^Q, then the test is merely "$f^Q(\widehat{\mathrm{w}}) = 0$?" We have cast the whole algorithm on $\mathbb{C}^N$, but it would be equivalent to cast it on complex projective space $\mathbb{P}^N$ and use Q as the hyperplane at infinity.

As a cautionary remark, note that the algorithm depends on A and B being complete solution sets of the given polynomial subsystems, excepting the same set Q. Specifically, test (a) is not valid if B is a partial list of components in $V(f^B) \setminus Q$. Indeed, suppose $B' \subset B$ is a partial set of components and we wish to find $A \cap B'$. Then, a point $\mathbf{w}_A$ for which $f^B(\mathbf{w}_a) = 0$ is necessarily in B but not necessarily in B', so it would be unjustified to pass it directly into the output witness set. A similar objection holds for test (b) if A were not the complete solution set.

3.3. Solving equation by equation. Limiting one of the subsystems to one single equation the subsystem-by-subsystem solver reduces to the paradigm of solving systems equation by equation. Accordingly, we begin by computing a witness set X^i for the solution set $V(f_i)$, $i = 1, 2, \ldots, n$ of each individual polynomial. If any polynomial is identically zero, we drop it and decrement n. If any polynomial is constant, we terminate immediately, returning a null result. Otherwise, we find $X^i = (V(f_i) \cap L) \setminus Q$, where L is a 1-dimensional generic affine linear subspace. A linear parameterization of L involves just one variable, so X^i can be found with any method for solving a polynomial in one variable, discarding any points that fall in Q.

Next, we randomly choose the affine linear subspaces that will cut out the witness sets for any lower dimensional components that appear in succeeding intersections.

The algorithm proceeds by setting $W^1 = X^1$ and then computing $W^{k+1} =$ **SysBySys** $(W^k, X^{k+1}; Q)$ for $k = 1, 2, \ldots, n-1$. The output of stage k is a collection of witness sets $W^{k+1}_{\langle i \rangle}$ for i in the range from 1 to $\min(N, k+1)$. (Recall that $\langle i \rangle$ in the subscript means codimension i.) Of course, some of these may be empty, in fact, in the case of a total intersection, only the lowest dimensional one, $W^{k+1}_{\langle k+1 \rangle}$, is nontrivial.

In applying the subsystem-by-subsystem method to this special case, we can streamline the flowchart a bit, due to the fact that $V(f_{k+1})$ is a hypersurface. The difference comes in the shortcuts that allow some witness points to avoid the diagonal homotopy.

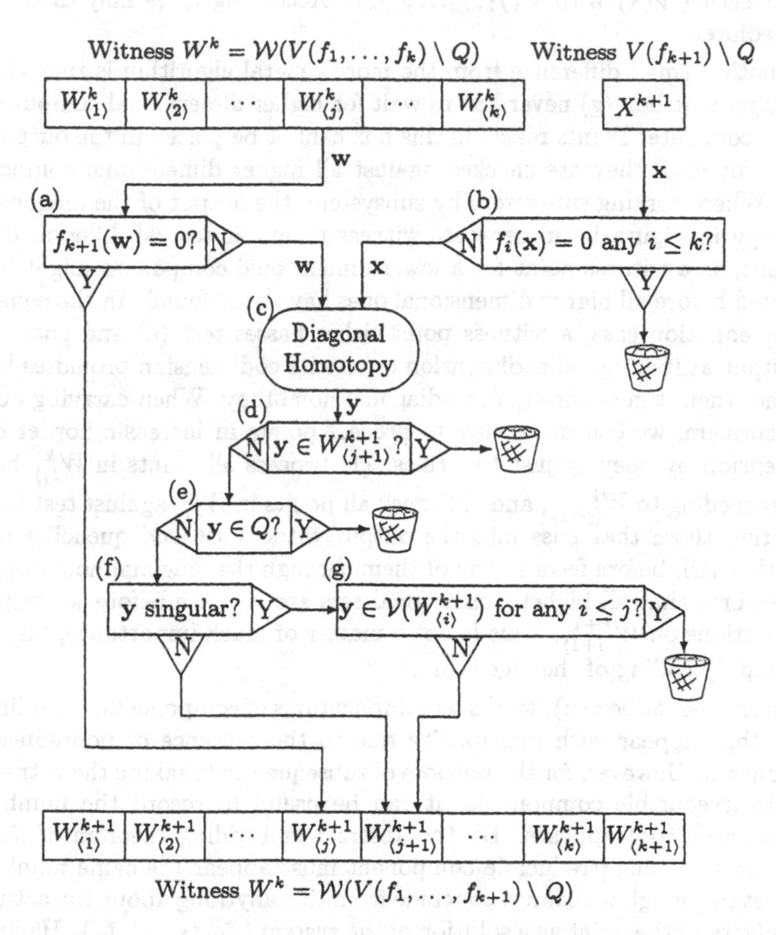

FIG. 2. *Stage k of equation-by-equation generation of witness sets for $V(f_1,\ldots,f_n) \in \mathbb{C}^N \setminus Q$. This figure is derived from Fig. 16.1 on page 295 of [29]: copyrighted 2005 by World Scientific Press.*

The first difference is at the output of test (a), which now sends $\mathbf{w}$ directly to the final output without any testing for duplicates. This is valid because we assume that on input $\mathcal{V}(\mathbf{w})$ is not contained within any higher dimensional component of $\mathcal{V}(W^k)$, and in the intersection with hypersurface $V(f_{k+1})$ that is the only place a duplication could have come from.

On the opposing side, test (b) is now stronger than before. The witness point $\mathbf{x}$ only has to satisfy one polynomial among $f_1, f_2, \ldots, f_k$ in order to receive special treatment. This is because we already threw out any polynomials that are identically zero, so if $f_j(\mathbf{x}) = 0$ it implies that $\mathcal{V}(\mathbf{x})$ is a factor of $V(f_j)$. But the intersection of that factor with all the other $V(f_i)$, $i \ne j$, is already in $V(f_1, f_2, \ldots, f_k)$, so nothing new can come out of intersecting $\mathcal{V}(\mathbf{x})$ with $V(f_1, f_2, \ldots, f_k)$. Accordingly, we may discard $\mathbf{x}$ immediately.

Another small difference from the more general algorithm is that the test for junk at box (g) never has to wait for higher dimensional computations to complete. Points reaching this box cannot be placed in the output witness set until they are checked against all higher dimensional components. When working subsystem by subsystem, the output of the diagonal homotopy in Figure 1 can generate witness points at several different dimensions, so a witness point for a lower dimensional component might be generated before all higher dimensional ones have been found. In the equation by equation case, a witness point either passes test (a) and goes to the output at its original codimension or else its codimension promotes by just one when it goes through the diagonal homotopy. When carrying out the algorithm, we can make sure to process points in increasing order of codimension by obeying just two rules: (1) process all points in $W^k_{\langle j \rangle}$ before proceeding to $W^k_{\langle j+1 \rangle}$, and (2) check all points in $W^k_{\langle j \rangle}$ against test (a), depositing those that pass into the output witness set and queueing up those that fail, before feeding any of them through the diagonal homotopy. This ensures that all higher dimensional sets are in place before we begin computations on $W^{k+1}_{\langle j+1 \rangle}$. This is not a matter of much importance, but it can simplify coding of the algorithm.

In the test at box (d), we discard duplications of components, including points that appear with multiplicity due to the presence of nonreduced components. However, for the purpose of subsequently breaking the witness set into irreducible components, it can be useful to record the number of times each root appears. By the abstract embedding theorem of [26], points on the same irreducible component must appear the same number times, even though we cannot conclude from this anything about the actual multiplicity of the point as a solution of the system $\{f_1, f_2, \ldots, f_n\}$. Having the points partially partitioned into subsets known to represent distinct components will speed up the decomposition phase.

A final minor point of efficiency is that if $n > N$, we may arrive at stage $k \geq N$ with some zero dimensional components, $W^k_{\langle N \rangle}$. These do not proceed to the diagonal homotopy: if such a point fails test (a), it is not a solution to system $\{f_1, f_2, \ldots, f_{k+1}\} = 0$, and it is discarded.

3.4. Seeking only nonsingular solutions. In the special case that $n \leq N$, we may seek only the multiplicity-one components of codimension n. For $n = N$, this means we seek only the nonsingular solutions of the system. In this case, we discard points that pass test (a), since they give higher dimensional components. Indeed, as discussed in [29, Chapter 12.2.1], for any $j \leq n$, any component Z of $V(f_1, \ldots, f_j)$ has codimension at most j, and if it is less than j, any point on Z will lie on a component of $V(f_1, \ldots, f_n)$ of codimension less than n.

Furthermore, we keep only the points that test (f) finds to be nonsingular and discard the singular ones. To see this, note that if in the end we want points z where the Jacobian matrix of f at z has rank n, then it follows that the rank at z of the Jacobian of $f_1, \ldots, f_j$ for $j \leq n$, must be j. This can greatly reduce the computation for some systems.

In this way, we may use the diagonal homotopy to compute nonsingular roots equation by equation. This performs differently than traditional approaches based on continuation, which solve the entire system all at once. In order to eliminate solution paths leading to infinity, these traditional approaches use multihomogeneous formulations or toric varieties to compactify $\mathbb{C}^N$. But this does not capture other kinds of structure that give rise to positive dimensional components. The equation-by-equation approach has the potential to expose some of these components early on, while the number of intrinsic variables is still small, and achieves efficiency by discarding them at an early stage. However, it does have the disadvantage of proceedin in multiple stages. For example, in the case that all solutions are finite and nonsingular, there is nothing to discard, and the equation-by-equation approach will be less efficient than a one-shot approach. However, many polynomial systems of practical interest have special structures, so the equation-by-equation approach may be commendable. It is too early to tell yet, as our experience applying this new algorithm on practical problems is very limited. Experiences with some simple examples are reported in the next section.

4. Computational experiments.

The diagonal homotopies are implemented in the software package PHCpack [30], also available in PHClab [8]. See [25] for a description of a recent upgrade of this package to deal with positive dimensional solution components.

4.1. An illustrative example.

The illustrative example (see Eq. 2.1 for the system) illustrates the gains made by our new solver. While our previous sequence of homotopies needed 197 paths to find all candidate witness points, the new approach shown in Figure 3 tracks just 13 paths. Many of the paths take shortcuts around the diagonal homotopies, and five paths that diverge to infinity in the first diagonal homotopy need no further consideration. It happens that none of the witness points generated by the diagonal homotopies is singular, so there is no need for membership testing.

On a 2.4Ghz Linux workstation, our previous approach [19] requires a total of 43.3 cpu seconds (39.9 cpu seconds for solving the top dimensional embedding and 3.4 cpu seconds to run the cascade of homotopies to find all candidate witness points). Our new approach takes slightly less than a second of cpu time. So for this example our new solver is 40 times faster.

4.2. Adjacent minors of a general 2-by-9 matrix.

In an application from algebraic statistics [2] (see also [10] for methods dedicated for

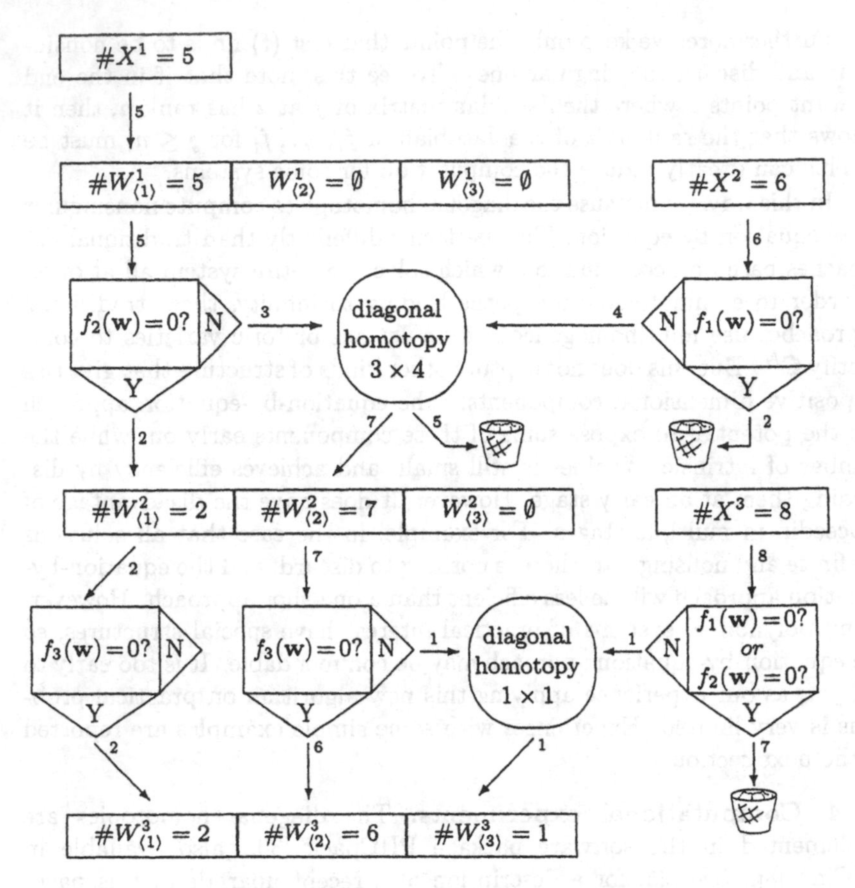

FIG. 3. *Flowchart for the illustrative example.*

these type of ideals) one considers all adjacent minors of a general matrix. For instance, consider this general 2-by-9 matrix:

$$\left[\begin{array}{ccccccccc} x_{11} & x_{12} & x_{13} & x_{14} & x_{15} & x_{16} & x_{17} & x_{18} & x_{19} \\ x_{21} & x_{22} & x_{23} & x_{24} & x_{25} & x_{26} & x_{27} & x_{28} & x_{29} \end{array} \right]$$

Two minors are adjacent if they share one neighboring column. Taking all adjacent minors from this general 2-by-9 matrix gives 8 quadrics in 18 unknowns. This defines a 10-dimensional surface, of degree 256.

We include this example to illustrate that the flow of timings is typical as in Table 2. Although we execute many homotopies, most of the work occurs in the last stage, because both the number of paths and the number of variables increases at each stage. We are using the intrinsic method of [27] to reduce the number of variables. With the older extrinsic method of [26], the total cpu time increases threefold from 158s to 502s.

TABLE 2

Timings on Apple PowerBook G4 1GHz for the 2 × 9 adjacent minors, a system of 8 quadrics in 18 unknowns.

stage	#paths			time/path	time
1	4	=	2 × 2	0.03s	0.11s
2	8	=	4 × 2	0.05s	0.41s
3	16	=	8 × 2	0.10s	1.61s
4	32	=	16 × 2	0.12s	3.75s
5	64	=	32 × 2	0.19s	12.41s
6	128	=	64 × 2	0.27s	34.89s
7	256	=	128 × 2	0.41s	104.22s
				total user cpu time	157.56s

4.3. A general 6-by-6 eigenvalue problem. Consider $f(\mathbf{x}, \lambda) = \lambda\mathbf{x} - A\mathbf{x} = \mathbf{0}$, where $A \in \mathbb{C}^{6\times 6}$, A is a random matrix. These 6 equations in 7 unknowns define a curve of degree 7, far less than what may be expected from the application of Bézout's theorem: $2^6 = 64$. Regarded as a polynomial system on $\mathbb{C}^7$, the solution set consists of seven lines, six of which are eigenvalue-eigenvector pairs while the seventh is the trivial line $\mathbf{x} = \mathbf{0}$.

Clearly, as a matter of practical computation, one would employ an off-the-shelf eigenvalue routine to solve this problem efficiently. Even with continuation, we could cast the problem on $\mathbb{P}^1 \times \mathbb{P}^6$ and solve it with a seven-path two-homogeneous formulation. However, for the sake of illustration, let us consider how the equation-by-equation approach performs, keeping in mind that the only information we use about the structure of the system is the degree of each equation. That is, we treat it just like any other system of 6 quadratics in 7 variables and let the equation-by-equation procedure numerically discover its special structure.

In a direct approach of solving the system in one total-degree homotopy, adding one generic linear equation to slice out an isolated point on each solution line, we would have 64 paths of which 57 diverge. This does not even consider the work that would be needed if we wanted to rigorously check for higher dimensional solution sets.

Table 3 shows the evolution of the number of solution paths tracked in each stage of the equation-by-equation approach. The size of each initial witness set is $\#(X^i) = 2$, so each new stage tracks two paths for every convergent path in the previous stage. If the quadratics were general, this would build up exponentially to 64 paths to track in the final stage, but the special structure of the eigenvalue equations causes there to be only $i + 2$ solutions at the end of stage i. Accordingly, there are only 12 paths to track in the final, most expensive stage, and only 40 paths tracked altogether. The seven convergent paths in the final stage give one witness point on each of the seven solution lines.

TABLE 3
Number of convergent and divergent paths on a general 6-by-6 eigenvalue problem.

stage in solver	1	2	3	4	5	total
#paths tracked	4	6	8	10	12	40
#divergent paths	1	2	3	4	5	15
#convergent paths	3	4	5	6	⑦	25

5. Conclusions. The recent invention of the diagonal homotopy allows one to compute intersections between algebraic sets represented numerically by witness sets. This opens up many new possibilities for ways to manipulate algebraic sets numerically. In particular, one may solve a system of polynomial equations by first solving subsets of the equations and then intersecting the results. We have presented a subsystem-by-subsystem algorithm based on this idea, which when carried to extreme gives an equation-by-equation algorithm. The approach can generate witness sets for all the solution components of a system, or it can be specialized to only seek the nonsingular solutions at the lowest dimension. Applying this latter form to a system of N equations in N variables, we come full circle in the sense that we are using methods developed to deal with higher dimensional solution sets as a means of finding just the isolated solutions.

Experiments with a few simple systems indicates that the method can be very effective. Using only the total degrees of the equations, the method numerically discovers some of their inherent structure in the early stages of the computation. These early stages are relatively cheap and they can sometimes eliminate much of the computation that would otherwise be incurred in the final stages.

In future work, we plan to exercise the approach on more challenging problems, especially ones where the equations have interrelationships that are not easily revealed just by examining the monomials that appear. Multihomogeneous homotopies and polyhedral homotopies are only able to take advantage of that sort of structure, while the equation-by-equation approach can reveal structure encoded in the coefficients of the polynomials. One avenue of further research could be to seek a formulation that uses multihomogeneous homotopies or polyhedral homotopies in an equation-by-equation style to get the best of both worlds.

We have not provided cost analyses of our algorithms. We consider this a challenging problem for future research. Even more important from our perspective would be tests of the actual performance on challenging problems, comparing our algorithms with symbolic approaches such as the algorithms of Lazard, Kalkbrenner, and Giusti and Heintz.

Acknowledgements. We acknowledge the assistance of Anton Leykin in constructing the dictionary shown in Table 1. We also thank T.-Y. Li helpful comments. We thank Yun Guan for her suggestions.

REFERENCES

[1] M. Beltrametti and A.J. Sommese. *The adjunction theory of complex projective varieties.* Expositions in Mathematics, **16**, Walter De Gruyter, Berlin, 1995.

[2] P. Diaconis, D. Eisenbud, and B. Sturmfels. Lattice Walks and Primary Decomposition. In *Mathematical Essays in Honor of Gian-Carlo Rota*, edited by B.E. Sagan, R.P. Stanley, Vol. **161** of *Progress in Mathematics*, pp. 173–193. Birkhäuser, 1998.

[3] A.V. Geramita. *The Curves Seminar at Queen's.* Vol. IX. Proceedings of the seminar held at Queen's University, Kingston, Ontario, 1992–1993. Queen's University Press, Ontario, Canada, 1993.

[4] M. Giusti and J. Heintz. La détermination de la dimension et des points isolées d'une variété algébrique peuvent s'effectuer en temps polynomial. In *Computational Algebraic Geometry and Commutative Algebra, Cortona 1991*, edited by D. Eisenbud and L. Robbiano, Symposia Mathematica XXXIV, pp. 216–256. Cambridge UP, 1993.

[5] M. Giusti and J. Heintz. Kronecker's smart, little black boxes. In *Foundations of Computational Mathematics* edited by DeVore, R.A. and Iserles, A. and Süli, E., Vol. **284** of *London Mathematical Society Lecture Note Series*, pp. 69–104. Cambridge University Press, 2001.

[6] M. Giusti, J. Heintz, J.E. Morais, and L.M. Pardo. When polynomial equation systems can be "solved" fast? In G. Cohen, M. Giusti, and T. Mora, editors, *Applied Algebra, Algebraic Algorithms and Error-Correcting Codes. 11th International Symposium, AAECC-11. Paris, France, July 1995*, Vol. **948** of *Lecture Notes in Computer Science*, pp. 205–231. Springer-Verlag, 1995.

[7] M. Giusti, G. Lecerf, and B. Salvy. A Gröbner free alternative for polynomial system solving. *Journal of Complexity* **17**(1):154–211, 2001.

[8] Y. Guan and J. Verschelde. PHClab: A MATLAB/Octave interface to PHCpack. Submitted for publication.

[9] J. Harris. *Curves in projective space*, with the collaboration of D. Eisenbud. Université de Montreal Press, Montreal (Québec), Canada, 1982.

[10] S. Hosten and J. Shapiro. Primary Decomposition of Lattice Basis Ideals. *J. of Symbolic Computation* **29**(4–5):625–639, 2000.

[11] B. Huber, F. Sottile, and B. Sturmfels. Numerical Schubert calculus. *J. of Symbolic Computation* **26**(6):767–788, 1998.

[12] B. Huber and J. Verschelde. Pieri homotopies for problems in enumerative geometry applied to pole placement in linear systems control. *SIAM J. Control Optim.* **38**(4):1265–1287, 2000.

[13] M. Kalkbrenner. A generalized Euclidean algorithm for computing triangular representations of algebraic varieties. *J. Symbolic Computation* **15**:143–167, 1993.

[14] D. Lazard. A new method for solving algebraic systems of positive dimension. *Discrete Appl. Math.* **33**:147–160, 1991.

[15] G. Lecerf. Computing the equidimensional decomposition of an algebraic closed set by means of lifting fibers. *Journal of Complexity* **19**(4):564–596, 2003.

[16] T.Y. Li. Numerical solution of polynomial systems by homotopy continuation methods. In *Handbook of Numerical Analysis. Vol. XI. Special Volume: Foundations of Computational Mathematics*, edited by F. Cucker, pp. 209–304. North-Holland, 2003.

[17] M.M. Maza, G. Reid, R. Scott, and W. Wu. On approximate triangular decompositions in dimension zero. *J. of Symbolic Computation* **42**(7):693–716, 2007.

[18] M.M. Maza, G. Reid, R. Scott, and W. Wu. On approximate linearized triangular decompositions. In *Symbolic-Numeric Computation*, edited by D. Wang and L. Zhi, pp. 279–298. Trends in Mathematics, Birkhäuser, 2007.

[19] A.J. SOMMESE AND J. VERSCHELDE. Numerical homotopies to compute generic points on positive dimensional algebraic sets. *Journal of Complexity* **16**(3): 572–602, 2000.

[20] A.J. SOMMESE, J. VERSCHELDE AND C.W. WAMPLER. Numerical decomposition of the solution sets of polynomial systems into irreducible components. *SIAM J. Numer. Anal.* **38**(6):2022–2046, 2001.

[21] A.J. SOMMESE, J. VERSCHELDE, AND C.W. WAMPLER. Numerical irreducible decomposition using projections from points on the components. In *Symbolic Computation: Solving Equations in Algebra, Geometry, and Engineering*, Vol. **286** of *Contemporary Mathematics*, edited by E.L. Green, S. Hoşten, R.C. Laubenbacher, and V. Powers, pp. 37–51. AMS 2001.

[22] A.J. SOMMESE, J. VERSCHELDE, AND C.W. WAMPLER. Using monodromy to decompose solution sets of polynomial systems into irreducible components. In *Application of Algebraic Geometry to Coding Theory, Physics and Computation*, edited by C. Ciliberto, F. Hirzebruch, R. Miranda, and M. Teicher. Proceedings of a NATO Conference, February 25–March 1, 2001, Eilat, Israel, pp. 297–315, Kluwer Academic Publishers.

[23] A.J. SOMMESE, J. VERSCHELDE, AND C.W. WAMPLER. Symmetric functions applied to decomposing solution sets of polynomial systems. *SIAM J. Numer. Anal.* **40**(6):2026–2046, 2002.

[24] A.J. SOMMESE, J. VERSCHELDE, AND C.W. WAMPLER. A method for tracking singular paths with application to the numerical irreducible decomposition. In *Algebraic Geometry, a Volume in Memory of Paolo Francia*, edited by M.C. Beltrametti, F. Catanese, C. Ciliberto, A. Lanteri, C. Pedrini, pp. 329–345, W. de Gruyter, 2002.

[25] A.J. SOMMESE, J. VERSCHELDE, AND C.W. WAMPLER. Numerical irreducible decomposition using PHCpack. In *Algebra, Geometry, and Software Systems*, edited by M. Joswig and N. Takayama, pp. 109–130, Springer-Verlag 2003.

[26] A.J. SOMMESE, J. VERSCHELDE, AND C.W. WAMPLER. Homotopies for Intersecting Solution Components of Polynomial Systems. *SIAM J. Numer. Anal.* **42**(4):1552–1571, 2004.

[27] A.J. SOMMESE, J. VERSCHELDE, AND C.W. WAMPLER. An intrinsic homotopy for intersecting algebraic varieties. *Journal of Complexity* **21**:593–608, 2005.

[28] A.J. SOMMESE AND C.W. WAMPLER. Numerical algebraic geometry. In *The Mathematics of Numerical Analysis*, edited by J. Renegar, M. Shub, and S. Smale, Vol. **32** of *Lectures in Applied Mathematics*, pp. 749–763, 1996. Proceedings of the AMS-SIAM Summer Seminar in Applied Mathematics, Park City, Utah, July 17-August 11, 1995, Park City, Utah.

[29] A.J. SOMMESE AND C.W. WAMPLER. *The Numerical solution of systems of polynomials arising in engineering and science.* World Scientific Press, Singapore, 2005.

[30] J. VERSCHELDE. Algorithm 795: PHCpack: A general-purpose solver for polynomial systems by homotopy continuation. *ACM Transactions on Mathematical Software* **25**(2):251–276, 1999. Software available at `http://www.math.uic.edu/~jan`.

LIST OF WORKSHOP PARTICIPANTS

- Douglas N. Arnold, Institute for Mathematics and its Applications, University of Minnesota Twin Cities
- Donald G. Aronson, Institute for Mathematics and its Applications, University of Minnesota Twin Cities
- Evgeniy Bart, Institute for Mathematics and its Applications, University of Minnesota Twin Cities
- Daniel J. Bates, Institute for Mathematics and its Applications, University of Minnesota Twin Cities
- Gian Mario Besana, Department of Computer Science-Telecommunications, DePaul University
- Laurent Buse, Project GALAAD, Institut National de Recherche en Informatique Automatique (INRIA)
- Eduardo Cattani, Department of Mathematics and Statistics, University of Massachusetts
- Ionut Ciocan-Fontanine, Institute for Mathematics and its Applications, University of Minnesota Twin Cities
- Maria Angelica Cueto, Departamento de Matematica - FCEyN, University of Buenos Aires
- Wolfram Decker, Fachrichtung Mathematik, Universität des Saarlandes
- Bernard Deconinck, Department of Applied Mathematics, University of Washington
- Jesus Antonio De Loera, Department of Mathematics, University of California
- Harm Derksen, Department of Mathematics, University of Michigan
- Alicia Dickenstein, Departamento de Matemática, Facultad de Ciencias Exactas y Naturales, Universidad de Buenos Aires
- Jintai Ding, Department of Mathematical Sciences University of Cincinnati
- Sandra Di Rocco, Department of Mathematics, Royal Institute of Technology (KTH)
- Xuan Vinh Doan, Operations Research Center, Massachusetts Institute of Technology
- Kenneth R. Driessel, Mathematics Department, Iowa State University
- David Eklund, Matematiska Institutionen, Royal Institute of Technology (KTH)
- Makan Fardad, Department of Electrical and Computer Engineering, University of Minnesota Twin Cities

- Xuhong Gao, Department of Mathematical Sciences, Clemson University
- Luis Garcia-Puente, Department of Mathematics, Texas A&M University
- Oleg Golubitsky, School of Computing, Queen's University
- Jason E. Gower, Institute for Mathematics and its Applications, University of Minnesota Twin Cities
- Genhua Guan, Department of Mathematics, Clemson University
- Marshall Hampton, Department of Mathematics and Statistics, University of Minnesota
- Gloria Haro Ortega, Institute for Mathematics and its Applications, University of Minnesota Twin Cities
- Michael Corin Harrison, Department of Mathematics and Statistics, University of Sydney
- Milena Hering, Institute for Mathematics and its Applications, University of Minnesota Twin Cities
- Christopher Hillar, Department of Mathematics, Texas A&M University
- Benjamin J. Howard, Institute for Mathematics and its Applications, University of Minnesota Twin Cities
- Evelyne Hubert, Project CAFE, Institut National de Recherche en Informatique Automatique (INRIA)
- Farhad Jafari, Department of Mathematics, University of Wyoming
- Itnuit Janovitz-Freireich, Department of Mathematics, North Carolina State University
- Anders Nedergaard Jensen, Institut for Matematiske Fag, Äarhus University
- Gabriela Jeronimo, Departamento de Matematica - FCEyN, University of Buenos Aires
- Roy Joshua, Department of Mathematics, Ohio State University
- Irina Kogan, Department of Mathematics, North Carolina State University
- Martin Kreuzer, Department of Mathematics, Universität Dortmund
- Teresa Krick, Departamento de Matematica - FCEyN, University of Buenos Aires
- Michael Kunte, Fachrichtung Mathematik, Universität des Saarlandes
- Song-Hwa Kwon, Institute for Mathematics and its Applications, University of Minnesota Twin Cities
- Oliver Labs, Mathematik und Informatik, Universität des Saarlandes
- Sanjay Lall, Department of Aeronautics and Astronautics, Stanford University

- Niels Lauritzen, Institut for Matematiske Fag, Äarhus University
- Anton Leykin, Institute for Mathematics and its Applications, University of Minnesota Twin Cities
- Tien-Yien Li, Department of Mathematics, Michigan State University
- Hstau Liao, Institute for Mathematics and its Applications, University of Minnesota Twin Cities
- Tie Luo, Division of Mathematical Sciences, National Science Foundation
- Gennady Lyubeznik, School of Mathematics, University of Minnesota Twin Cities
- Diane Maclagan, Department of Mathematics, Rutgers University
- Susan Margulies, Department of Computer Science, University of California
- Hannah Markwig, Institute for Mathematics and its Applications, University of Minnesota Twin Cities
- Thomas Markwig, Department of Mathematics, Universität Kaiserslautern
- Guillermo Matera, Instituto de Desarrollo Humano, Universidad Nacional de General Sarmiento
- Laura Felicia Matusevich, Department of Mathematics, Texas A&M University
- Richard Moeckel, School of Mathematics, University of Minnesota Twin Cities
- Bernard Mourrain, Project GALAAD, Institut National de Recherche en Informatique Automatique (INRIA)
- Uwe Nagel, Department of Mathematics, University of Kentucky
- Jiawang Nie, Institute of Mathematics and its Applications, University of Minnesota Twin Cities
- Michael E. O'Sullivan, Department of Mathematics and Statistics, San Diego State University
- Jang-Woo Park, Department of Mathematical Sciences, Clemson University
- Pablo A. Parrilo, Laboratory for Information and Decision Systems, Massachusetts Institute of Technology
- Matt Patterson, Department of Applied Mathematics, University of Washington
- Paul Pedersen, Department of Computer and Computation Science, Los Alamos National Laboratory
- Chris Peterson, Department of Mathematics, Colorado State University
- Sonja Petrovic, Department of Mathematics, University of Kentucky
- Ragni Piene, Centre of Mathematics for Applications, University of Oslo

- Sorin Popescu, Department of Mathematics, SUNY
- Kristian Ranestad, Department of Mathematics, University of Oslo
- Gregory J. Reid, Department of Applied Mathematics, University of Western Ontario
- Victor Reiner, School of Mathematics, University of Minnesota Twin Cities
- Joel Roberts, School of Mathematics, University of Minnesota Twin Cities
- J. Maurice Rojas, Department of Mathematics, Texas A&M University
- Fabrice Rouillier, Projet SALSA, Institut National de Recherche en Informatique Automatique (INRIA)
- David Rusin, Department of Mathematical Sciences, Northern Illinois University
- Michael Sagraloff, Algorithms and Complexity, Max-Planck-Institut für Informatik
- Bruno Salvy, Projet ALGO, Institut National de Recherche en Informatique Automatique (INRIA)
- Arnd Scheel, Institute for Mathematics and its Applications, University of Minnesota Twin Cities
- Eric Schost, LIX, École Polytechnique
- Frank-Olaf Schreyer, Mathematik und Informatik, Universität des Saarlandes
- Chehrzad Shakiban, Institute of Mathematics and its Applications, University of Minnesota Twin Cities
- Tanush Shaska, Department of Mathematics and Statistics, Oakland University
- Andrew J. Sommese, Department of Mathematics, University of Notre Dame
- Ivan Soprunov, Department of Mathematics, Cleveland State University
- Frank Sottile, Department of Mathematics, Texas A&M University
- Steven Sperber, School of Mathematics, University of Minnesota Twin Cities
- Dumitru Stamate, School of Mathematics, University of Minnesota Twin Cities
- Michael E. Stillman, Department of Mathematics, Cornell University
- Bernd Sturmfels, Department of Mathematics, University of California at Berkeley
- David Swinarski, Department of Mathematics, Columbia University
- Agnes Szanto, Department of Mathematics, North Carolina State University
- Nobuki Takayama, Department of Mathematics, Kobe University

- Kathleen Iwancio Thompson, North Carolina State University
- Carl Toews, Institute for Mathematics and its Applications, University of Minnesota Twin Cities
- Ravi Vakil, Department of Mathematics, Stanford University
- Mark van Hoeij, Department of Mathematics, Florida State University
- Mauricio Velasco, Department of Mathematics, Cornell University
- Jan Verschelde, Department of Mathematics, Statistics and Computer Science, University of Illinois at Chicago
- John Voight, Institute for Mathematics and its Applications, University of Minnesota Twin Cities
- Hans-Christian von Bothmer, Institut für Mathematik, Universität Hannover
- Charles W. Wampler, General Motors Research and Development Center
- Daqing Wan, Department of Mathematics, University of California
- Mingsheng Wang, Institute of Software, Chinese Academy of Sciences
- Oliver Wienand, Department of Mathematics, University of California
- Wenyuan Wu, Department of Applied Mathematics, University of Western Ontario
- Ruriko Yoshida, Department of Statistics, University of Kentucky
- Cornelia Yuen, Department of Mathematics, University of Kentucky
- Zhonggang Zeng, Department of Mathematics, Northeastern Illinois University
- Mingfu Zhu, Department of Mathematical Sciences, Clemson University

1997–1998 Emerging Applications of Dynamical Systems
1998–1999 Mathematics in Biology
1999–2000 Reactive Flows and Transport Phenomena
2000–2001 Mathematics in Multimedia
2001–2002 Mathematics in the Geosciences
2002–2003 Optimization
2003–2004 Probability and Statistics in Complex Systems: Genomics, Networks, and Financial Engineering
2004–2005 Mathematics of Materials and Macromolecules: Multiple Scales, Disorder, and Singularities
2005-2006 Imaging
2006-2007 Applications of Algebraic Geometry
2007-2008 Mathematics of Molecular and Cellular Biology
2008-2009 Mathematics and Chemistry

IMA SUMMER PROGRAMS

1987 Robotics
1988 Signal Processing
1989 Robust Statistics and Diagnostics
1990 Radar and Sonar (June 18–29)
New Directions in Time Series Analysis (July 2–27)
1991 Semiconductors
1992 Environmental Studies: Mathematical, Computational, and Statistical Analysis
1993 Modeling, Mesh Generation, and Adaptive Numerical Methods for Partial Differential Equations
1994 Molecular Biology
1995 Large Scale Optimizations with Applications to Inverse Problems, Optimal Control and Design, and Molecular and Structural Optimization
1996 Emerging Applications of Number Theory (July 15–26)
Theory of Random Sets (August 22–24)
1997 Statistics in the Health Sciences
1998 Coding and Cryptography (July 6–18)
Mathematical Modeling in Industry (July 22–31)
1999 Codes, Systems, and Graphical Models (August 2–13, 1999)
2000 Mathematical Modeling in Industry: A Workshop for Graduate Students (July 19–28)
2001 Geometric Methods in Inverse Problems and PDE Control (July 16–27)
2002 Special Functions in the Digital Age (July 22–August 2)

2003 Probability and Partial Differential Equations in Modern Applied Mathematics (July 21–August 1)
2004 n-Categories: Foundations and Applications (June 7–18)
2005 Wireless Communications (June 22–July 1)
2006 Symmetries and Overdetermined Systems of Partial Differential Equations (July 17–August 4)
2007 Classical and Quantum Approaches in Molecular Modeling (July 23-August 3)
2008 Geometrical Singularities and Singular Geometries (July 14-25)

IMA "HOT TOPICS" WORKSHOPS

- Challenges and Opportunities in Genomics: Production, Storage, Mining and Use, April 24–27, 1999
- Decision Making Under Uncertainty: Energy and Environmental Models, July 20–24, 1999
- Analysis and Modeling of Optical Devices, September 9–10, 1999
- Decision Making under Uncertainty: Assessment of the Reliability of Mathematical Models, September 16–17, 1999
- Scaling Phenomena in Communication Networks, October 22–24, 1999
- Text Mining, April 17–18, 2000
- Mathematical Challenges in Global Positioning Systems (GPS), August 16–18, 2000
- Modeling and Analysis of Noise in Integrated Circuits and Systems, August 29–30, 2000
- Mathematics of the Internet: E-Auction and Markets, December 3–5, 2000
- Analysis and Modeling of Industrial Jetting Processes, January 10–13, 2001
- Special Workshop: Mathematical Opportunities in Large-Scale Network Dynamics, August 6–7, 2001
- Wireless Networks, August 8–10 2001
- Numerical Relativity, June 24–29, 2002
- Operational Modeling and Biodefense: Problems, Techniques, and Opportunities, September 28, 2002
- Data-driven Control and Optimization, December 4–6, 2002
- Agent Based Modeling and Simulation, November 3–6, 2003
- Enhancing the Search of Mathematics, April 26-27, 2004
- Compatible Spatial Discretizations for Partial Differential Equations, May 11-15, 2004
- Adaptive Sensing and Multimode Data Inversion, June 27–30, 2004
- Mixed Integer Programming, July 25–29, 2005
- New Directions in Probability Theory, August 5–6, 2005
- Negative Index Materials, October 2-4, 2006

- The Evolution of Mathematical Communication in the Age of Digital Libraries, December 8-9, 2006
- Math is Cool! and Who Wants to Be a Mathematician?, November 3, 2006
- Special Workshop: Blackwell-Tapia Conference, November 3-4, 2006
- Stochastic Models for Intracellular Reaction Networks, May 11-13, 2008

SPRINGER LECTURE NOTES FROM THE IMA:

The Mathematics and Physics of Disordered Media
Editors: Barry Hughes and Barry Ninham
(Lecture Notes in Math., Volume 1035, 1983)

Orienting Polymers
Editor: J.L. Ericksen
(Lecture Notes in Math., Volume 1063, 1984)

New Perspectives in Thermodynamics
Editor: James Serrin
(Springer-Verlag, 1986)

Models of Economic Dynamics
Editor: Hugo Sonnenschein
(Lecture Notes in Econ., Volume 264, 1986)

IMA VOLUMES

Volume 142: Compatible Spatial Discretizations
Editors: Douglas N. Arnold, Pavel B. Bochev, Richard B. Lehoucq, Roy A. Nicolaides, and Mikhail Shashkov

Volume 143: Wireless Communications
Editors: Prathima Agrawal, Daniel Matthew Andrews, Philip J. Fleming, George Yin, and Lisa Zhang

Volume 144: Symmetries and Overdetermined Systems of Partial Differential Equations
Editors: Michael Eastwood and Willard Miller, Jr.

Volume 145: Topics in Stochastic Analysis and Nonparametric Estimation
Editors: Pao-Liu Chow, Boris Mordukhovich, and George Yin

Volume 146: Algorithms in Algebraic Geometry
Editors: Alicia Dickenstein, Frank-Olaf Schreyer, and Andrew J. Sommese

Volume 147: Symmetric Functionals on Random Matrices and Random Matchings Problems by Grzegorz A. Rempała and Jacek Wesołowski

The full list of IMA books can be found at the website of Institute for Mathematics and its Applications: http://www.ima.umn.edu/springer/volumes.htm